Urban Flood Management

Urban Flood Management

Edited by
A. Szöllösi-Nagy

UNESCO Natural Sciences Sector, Paris, France

C. Zevenbergen

Dura Vermeer Business Development BV, Hoofddorp, The Netherlands

A.A. BALKEMA PUBLISHERS   Leiden/London/New York/Philadelphia/Singapore

The **First International Expert Meeting on Urban Flood Management**, which was held in Rotterdam, The Netherlands, on November 20 and 21, 2003, was organised by: the Dutch Ministry of Transport, Public Works and Water Management, The Netherlands Water Partnership (NWP), The International Network on Spatial Development and Water (Spid'O), and Aqua Delta Forum (ADF). The meeting was sponsored by the Dutch Ministry of Transport, Public Works and Water Management, the Dutch Ministry of Housing, Spatial Planning and Environment and The Netherlands Water Partnership (NWP).

Published by: A.A. Balkema Publishers, a member of Taylor & Francis Group plc
www.balkema.nl and www.tandf.co.uk

ISBN 04 1535 998 8

Printed in Great-Britain

Table of Contents

Preface

In recent years a trend shift in flood management has become evident internationally. Whereas the classical approach was a technical one, nowadays "space for water" has become the guiding principle. Along similar lines, it has become increasingly apparent that urban flood management requires a combination of spatial and technical approaches in order to be effective.

With this trend shift, water management enters the world of land use policies: "water is seeping onto the land"! Different worlds, unaware of each other's existence, are forced to meet. Decisionmakers, policy strategists, urban planners, technicians and scientists experience serious communication problems. Whereas technicians in the water sector tend to use *calculation models*, spatial planners use *crayons* to draw up spatial land-use plans. Whereas decisionmakers in the field of urban planning policies are used to making *integrated* decisions, water managers tend to focus on the *sectoral* importance of flood protection. Cost-benefit analysts are confronted with the problem of having to express costs and benefits of the "quality of space" in quantitative terms in order to facilitate a comparison with the costs and benefits of a dyke as an anti-flood strategy. But rather than talking in terms of conflicts and communication problems, I believe that new perspectives and opportunities can be identified. The private sector is already investigating the opportunities to integrate water into their project development plans. Dutch farmers in the *Overdiepse Polder* presented a plan to combine space for the river and continue their agricultural activities. This plan has recently been adopted by the government and will now be put to practice.

Therefore, we need to integrate these worlds to find and realise space for the water. At the same time, innovative solutions for the claimed available and often scarce space need to be found by means of multifunctional land use. And finally, in order to realise an improvement in the quality of the urban environment.

I call upon you to listen to your new colleagues with an open mind. This is the first necessary step on the way to establishing an integration of the hitherto different worlds, eventually leading to effective and realistic urban flood management measures!

The Director-General for Spatial Planning in The Netherlands,

Ineke Bakker

List of Contributors

P.J.A. Baan, MSc.
WL/Delft Hydraulics
Rotterdamseweg 185
2629 HD Delft
The Netherlands

Prof. D. Crichton
The Benfield Hazard Research Centre
UCL
136 Gower Street (Lewis Building)
London, WC1E 6BT
United Kingdom

J.L. Fiselier, MSc.
DHV Group Amersfoort
Laan 1914 35
3818 EX Amersfoort
The Netherlands

Dr. S.L. Garvin
Building Research Establishment (BRE)
Kelvin Road
East Kilbride
Glasgow G75 0RZ
Scotland
United Kingdom

T.R. Geisler, MSc.
Institute of River & Coastal Engineering
Hamburg University of Technology
Denickestr. 22
D-21073 Hamburg
Germany

Dr. N.L. Miller
Atmospheric and Oceanic Sciences Group
Earth Sciences Division, Berkeley National Laboratory
90-1116 One Cyclotron Drive, Berkeley
CA 94720
USA

W. Oosterberg, MSc.
RIZA Institute for Inland Water Management and Waste Water Treatment
Postbus 17
8200 AA Lelystad
The Netherlands

Prof. Dr. E. Pasche
Technical University of Hamburg
River & Coastal Engineering
Denickestr. 22
D-21073 Hamburg
Germany

Prof. C.E.M. Tucci
Institute of Hydraulic Research, Federal University of Rio Grande do Sul
Caixa Postal 15029
91501-970
Porto Alegre – RS
Brazil

J. Verhagen, MSc.
Disaster Mitigation Institute (DMI)
A–51 (First Floor)
Defence Colony
New Delhi 110 024
India

Dr. A.O.N. Villanueva
Institute of Hydraulic Research
Federal University of Rio Grande do Sul
Caixa Postal 15029
91501-970
Porto Alegre – RS
Brazil

Prof. A. Szöllösi-Nagy, Ph.D.
UNESCO Natural Sciences Sector
1, rue de Miollis
75015 Paris
France

Dr. C. Zevenbergen
Dura Vermeer Business Development BV
P.O. Box 3098
2130 KB Hoofddorp
The Netherlands

1

Introduction

Andras Szöllösi-Nagy
UNESCO Natural Sciences Sector, Paris, France

Chris Zevenbergen
Dura Vermeer Business Development BV, Hoofddorp, The Netherlands

1.1   URBAN FLOODS

Floods and droughts have always been the natural disasters causing the greatest loss of human life and economic damage. As a result of an increase in vulnerability, flood impact statistics show an exponential growth in the number of flood events and affected populations. The greatest potential for flood disasters exists in the populous urban and peri-urban settlements of the globe. The characteristics of flood disasters evolve with the process of urbanisation and industrialisation. Gradually more and more potential flood disaster areas are developing. Some urban areas, however, are more vulnerable to floods than others, depending on both the sensitivity to floods and the ability to absorb the flood impact.

The magnitude of the flood risk depends on a combination of natural factors influencing the frequency of floods and human-related factors influencing both flood frequency *and* flood impact. The increase in vulnerability of cities to flood disasters arises predominantly from the following factors (Gladwell and Sim, 1993):

– systematic degradation of natural ecosystems;
– increased urban migration;
– unplanned occupation and unsustainable planning and building practices.

It is increasingly recognised that apart from these factors, climate change will intensify the already significant threat posed by flood disasters to human settlements, most particularly in the rapid urbanising third world. A striking transformation can be seen in the countries in this area, where unplanned urbanisation and poverty are dramatically increasing vulnerability to floods.

Urbanisation is a process that has multifaceted effects on the hydrology of a catchment. For instance, increased urbanisation of many cities in delta regions has caused considerable portions of cities to subside due to ground water abstraction. As a direct consequence, land subsidence will exacerbate the urban flood hazard and the impact of flooding. Catchment urbanisation also leads to the replacement of previously natural surfaces with impermeable ones, a reduction of soil infiltration and an increase in run-off volumes. Urban drainage is increasingly viewed as the "Achilles heel" of flood management systems (Ashley, 2003). Urban flood management strategies have traditionally focussed on water originating in the city itself and on protection against local flooding through the provision of a hydraulically efficient drainage system to accommodate the expected increase in storm water discharge (Maksimovic, 2001).

1.2   FLOODS AND DAMAGE

Since 1990 devastating floods were registered in virtually every part of the world, e.g., Mozambique, India, Central America, China, Poland, Germany and the Czech Republic. In this period, the material losses of 30 floods exceeded one billion US$ and/or the number of fatalities was greater than 1000. According to MunichRe (2001), floods accounted for 50% of economic losses caused by natural disasters in 2000. Most of the economic losses are sustained in the developed and industrialised countries, but expressed in relative terms (losses by percentage of GDP) it is the developing countries which show the highest losses. The opposite is the case for the number of people affected (fatalities, homelessness, diseases). During the last decades the reported number of disasters caused by floods in Europe has dramatically increased, from 31 in the period 1973–1982 to 179 during the last decade (see Figure 1.1). In the entire period floods caused a total of 264 disasters. According to Hoyois and Guha-Sapir (2003) the total amount of reported damage from disastrous floods in Europe increases from 5.7 billion in the first to 48.6 billion Euro in the last

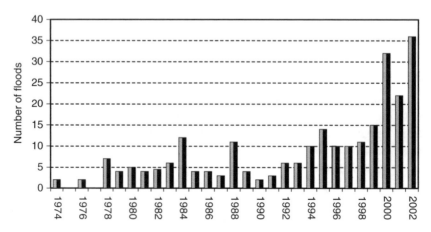

Figure 1.1. Total number of floods reported in Europe (modified after Hoyois and Guha-Sapir (2003).

three decades. The sum total of direct economic loss in Europe in this entire period amounted to 72 billion Euro.

The impact of floods on cities can devastate national economies and industrial markets at an international level. Due to urban concentration of the population the greatest potential for flood disasters exists in the most populous cities. For instance, in Japan where about 50% of the population and about 75% of the assets concentrate in urban agglomerates on flood plains. It is important at this point to note that in Japan over the last three decades, the number of flood disasters and the amount of economic loss has increased, while the number of fatalities has decreased. This decrease is due to improved early warning systems and increased preparedness, among other factors. An illustration of this trend is the most severe flood disaster, which hit the Tokai area with 2.1 million residents in September 2000. This flood caused 10 deaths and 978.3 billion Japanese yen in direct economic losses. As noted earlier, however, it is expected that in developing countries population growth and increased urban migration will further increase the number of deaths and the economic losses in cities, where nowadays more than 90% of natural disaster related deaths occur.

1.3   DEVELOPMENTS IN FLOOD MANAGEMENT

The way in which flood management is approached has evolved over time. Historically, there have been four successive approaches to flood hazard management (Green et al., 2000):

– Indigenous flood adaptation: communities have occupied flood-prone areas for many generations. Local adaptations develop to make them more resilient to floods: houses are constructed on stilts (e.g., in New Guinea, Benin, Thailand) and on higher ground and mounds (e.g., in The Netherlands, Bangladesh) to raise them above anticipated flood levels. In Bangladesh houses are even temporarily dismantled in times of high water. There are numerous examples of indigenous adaptations to floods, however, mainly due to rapid urbanisation and economic growth in conjunction with a growing faith in structural measures to protect communities from floods, these approaches have been abandoned. It goes without saying that consideration of these approaches may still prove useful in seeking a solution for today's problems.

– Flood control and defence: structural measures, such as flood embankments, dikes and flood control dams, were developed in the 19th and most of the 20th century. These large-scale engineering measures appeared to be very effective in controlling rivers and preventing flood water entering communities located in flood-prone areas. However, this second generation approach received much criticism in the late 20th century. Provision of flood protection, either in an urban or rural environment has led to cyclical patterns of investment – in the absence of land-use controls, the population density in former flood plains has increased, property values have risen and the pressure to provide a greater degree of flood protection has intensified. The effectiveness of physical flood control measures may change in time depending on the level of maintenance and on physical processes that affect sustainability, for example, changes in river morphology and sedimentation. Even in most European countries with resources for maintenance and more sophisticated flood warning systems, the physical flood protection infrastructure has on occasion failed to provide full protection.

– Non-structural approaches: as an alternative to the engineering solutions non-structural approaches have received much attention in the US and Western Europe. In this approach emphasis is laid on both the behaviour of people and on strategies which influence their behaviour (moving people away from floods, evacuating them using warning systems, etc.) and on land-use planning as a steering instrument to design and plan communities in safe areas or to better adapt them to the impacts of floods. Although public perception may be that physical intervention is sufficient, the reality is that considerable management effort is required and additional attention and resources need to be allocated for non-structural approaches to flood management (flood forecasting, land-use zoning, flood proofing, disaster preparedness, flood insurance) either in parallel or independent of structural forms of flood protection.

– Living with floods: the recurrence of disastrous floods have shown that many structural and non-structural strategies have failed to be effective. New options emerge, in which flood-prone areas are kept free from urbanisation and transformed into "green belts", through which flood waters can pass and temporarily be stored to prevent surrounding urbanised areas from flood damage. Simultaneously, these open areas may serve recreational purposes. It is becoming increasingly recognised that there is no single effective strategy, but a variety of strategies based on a holistic approach taking into account both the causes and the impact of floods and flood disasters on the catchment as a whole. The choice of measures to be implemented may also be dominated by local economics and social factors. The level of protection a city or region can afford and are willing to accept remains a principle question and imposes complex constraints, which cannot be ignored. A holistic approach thus considers the catchment as a unit for planning and management, in which sustainable water management is the guiding principle in the planning process.

1.4   URBAN FLOOD MANAGEMENT

Urban floods may have wide impact on natural, economic, social and cultural systems. Urban flood management approaches that fail to enhance the capacity

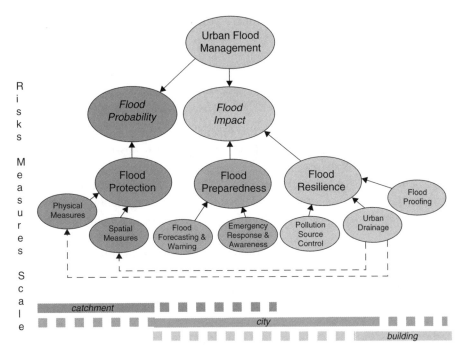

Figure 1.2. Urban flood management: measures.

of individuals and society to cope with floods are, therefore, likely to be counter productive in the longer term (IRMA, 2001). It is increasingly recognised that urban flood management calls for a holistic approach as outlined in the previous section, considering measures to reduce flood probability and flood impact (see Figure 1.2). The selection of measures including flood protection, flood preparedness and flood mitigation should address different spatial scales and support sustainable development of the entire catchment, while maximising the economic efficiency of land use. The latter implies that, for instance, urban drainage solutions should be at best integrated into flood management plans for the entire catchment (Maksimovic, 2001). Measures to reduce flood impact aim to reduce the risk of urban flooding at a *local* scale using a portfolio of adaptation strategies. These strategies are directed at increasing flood resilience of the built environment by adopting sustainable planning, pollution source control, urban drainage and flood proof building practices.

Measures to increase flood resilience are developed with the objective of reducing the impact of flooding on an urban community. Financial benefit in terms of flood damage reduction is an important criterion for the evaluation, ranking and selection of appropriate options (Price and McInally, 2001). However, the lack of data and analysis of the technical performance and the economics of implementation and maintenance hampers promotion of flood proofing and urban drainage systems.

1.5   FLOOD PROOFING

Flood-proofing techniques comprise an arsenal of engineering options to protect buildings and infrastructure from flood damage and to keep flood water away from damageable properties.

Flood-proofing techniques have been developed to the greatest extent in the US. Some of these techniques originate in indigenous flood adaptation approaches. Two categories of flood-proofing techniques can be distinguished (Green et al., 2000):
– permanent flood proofing; the building is constructed or transformed so as to reduce flood damage;
– contingent flood proofing; action is taken (e.g., installation of barriers across all openings) immediately preceding the onset of the flood.

Flood-proofing techniques have been widely applied specifically to protect historical buildings from flood damage and to allow low density urban development in flood prone areas. Its application, however, has been restricted to the developed world only. With the growing pressure to obtain urban land due to population growth and rural-urban migration, flood-proofing solutions may introduce a new and valuable option to city development projects located in risk areas in developing countries. Reducing the cost of flood proofing housing, however, using new or improved construction techniques, the application of low cost building technology and the use of appropriate materials is one of the prerequisites for the introduction of flood-proofing techniques in developing countries.

The following are examples of permanent flood-proofing techniques (Green et al., 2000):
– Buildings on fill and artificial mounds: buildings are constructed on fill or artificial mounds raised above the design flood level. This method does not require design modifications, and if the design flood water level is exceeded, the depth of water over the fill will be shallow and of short duration. It is not, however, a practical alternative for protecting existing buildings.

– Floating and amphibious buildings: floating buildings may provide an alternative to elevated buildings in permanently inundated areas (e.g., detention basins). These houses can tolerate a fluctuating water level and are therefore well protected against flood water. Floods and subsequent dike reinforcements in the catchment basin have led to the development of amphibious houses in The Netherlands: i.e., houses that float during floods. To enable the houses to move with the water level, they are built on floating concrete bodies with a coupling construction. At low water levels, the houses rest on a concrete foundation. They have a wooden frame construction in order to keep them as light as possible. To prevent the houses from floating off during a flood, they are anchored to flexible mooring posts. A difference in water level of up to about 5 metres can be accommodated.

– Buildings on piers, piles, columns or bearing walls: for many generations elevated houses have been constructed in delta areas and along rivers. Elevating structures may provide reliable protection against flood damage. This method uses land efficiently, does not raise the flood level, and has minimal adverse effects on flood flows. Examples of recent, high density developments using elevated structures can be found in Japan (Tsurumi river basin) and the UK (Thames Gate).

– Closure and seal techniques: these techniques are typical examples of "dry flood proofing". The walls of buildings are sealed to make them impermeable to flood water. Flood proofing the lower levels of buildings by seal techniques requires that they are strong enough to withstand cracking from the lateral and uplift pressure of the water. Accordingly, careful design of drainage systems, floor slabs, basement walls, lower windows and all entrances is essential. This method can be used for existing structures if they are of adequate strength and built on soils of low permeability. The method requires an adequate flood warning system and pre-planned evacuation measures, since there is greater risk of catastrophic damage if the design flood level is exceeded. It is not suitable for floods of long duration or where high flood depths are possible.

– Wet flood-proofing: as opposed to dry flood-proofing with this technique it is accepted that the interior of the building will be flooded. Flood damage is kept to a minimum by using special water-resistant construction materials in the lower levels of the building. Wet flood-proofing is frequently the only method of controlling or reducing flood damage to existing buildings in areas subject to flooding. It is seldom advocated for new buildings because of the cost of drying out, the delay in return to use and the clean-up required after a flood.

1.6   URBAN DRAINAGE SYSTEMS

Urbanisation reduces the amount of water infiltrating into the ground and increases surface run-off. Urban areas need, therefore, to be drained to remove surface water. In modern societies this has been implemented for some decades using underground pipe systems designed for quantity, preventing flooding locally by conveying the water away as quickly as possible. In conventional drainage systems this water is combined with urban waste water (e.g., foul water from toilets),

and drained through one combined sewer. Recent practice has separated drainage systems to provide separate sewers for the foul and the surface water. Simultaneously, urban drainage systems are more frequently transformed into integrated approaches for surface water drainage methods, which take into account the quantity, quality and amenity issues (Maksimovic and Tejada-Guibert, 2001). These new generation systems (Sustainable Urban Drainage Systems (SUDS)) are more sustainable than the conventional ones due to factors outlined below (CIRIA Report C521, 2000; Maksimovic, 2001):
– management of runoff flow rates;
– reduction of the impact of urbanisation on flooding;
– protection or enhancement of water quality;
– provision of a habitat for wildlife in urban watercourses;
– encouragement of natural groundwater recharge (where appropriate).

Urban drainage systems comprise a wide spectrum of techniques, which can be categorised in:
– permeable surfaces (gravel surfaces, porous blocks etc.);
– filter strips and swales;
– infiltration devices;
– basins and ponds.

At present there is still a lack of reliable information about the performance of SUDS and considerable concerns that these systems may not perform as expected, particularly during extreme rainfall events. There is also concern about the impact of run-off infiltration on the quality of groundwater. These and other concerns about the performance of SUDS currently form a barrier to their wide spread adoption. In developing countries it is increasingly recognised that urban drainage systems cannot be ignored and should be integrated into urban master plans. Transferring technologies and approaches which are appropriate in developed countries are, however, in most cases inappropriate in developing countries.

1.7 ABOUT THIS BOOK

Given the ongoing process of urbanisation and climate change, urban floods are becoming a key focus for those who are engaged with the future of cities around the world. The aim of this book is to discuss some of the recent developments in this broad field of urban flood management. This introductory chapter has described the general context of this emerging and multifaceted topic. It summarises the evolution in approaches which have been developed to cope with floods and flood disasters in general, outlines an holistic approach to urban flood management and elaborates on strategies which aim at increasing flood resilience of the built environment. The following chapters contain papers which were presented at the first International Expert Meeting on Urban Flood Management, which was held in Rotterdam, The Netherlands, in November 2003. Chapters 2, 3, 4 and 5 will present causes and impacts of urban floods and practical experiences with land-use and urban floods in Brazil, India and Europe. An analysis of risk perception and preparedness in The Netherlands and experiences with flood insurance in general are presented in Chapters 6 and 7. It has been argued that in evaluating risks, governments and experts apply an analytical approach emphasising the use of analytical data, while the views of the public are based on experiences and emotions. Flood risk management strategies should, therefore, be developed in close co-operation with the citizens involved. Furthermore, while the demand for flood insurance increases, insurers need to utilise incentives for risk reducing behaviour, so that efforts to mitigate flood damage are encouraged. Chapters 8 and 9 will focus on flood mitigation strategies with the emphasis on flood resilience in the built environment. It concludes that there is a lack of information on the possibility of improving the flood resistance of buildings in general and on the subsequent economic benefits. Wet-proofing and dry-proofing techniques may provide a huge potential in the reduction of flood damage. These techniques, however, are still in an early stage of their technological development. Since flood modelling has been suggested as one of the most effective tools to guide the planning of a more resilient infrastructure in the future, a state of the art review of climate change and hydrology modelling is presented in Chapter 10. Chapter 11 offers concluding remarks and suggestions for research into the improvement of the performance of urban flood management strategies in general.

REFERENCES

Building Research Establishment. 1996. *Design Guidance on Flood Damage to Dwellings*, (ISBN 0 11 495776 2) Building Research Establishment, East Kilbride, Scotland.

Price, D.J. and McInally, G. 2001: *Climate Change: Review of Levels of Protection Offered by Flood Prevention Schemes*. Scottish Executive Central Research Unit Report No 12. Edinburgh, May 2001.

IRMA. 2002. *Non Structural Flood Plain Management*.

Institute for Catastrophic Loss Reduction, Ontario, Canada. October 2001. *Proceedings of a conference on non structural solutions to water problems*.

River Restoration Centre. *Manual of River Restoration Techniques*. February 1999. 100pp.

CIRIA Report C521. London 2000. *Sustainable urban drainage systems. Design manual for Scotland and Northern Ireland*.

Green, C., Parker, J.D. and Tunstall, S.M. November 2000. *Assessment of flood control and management options, World Commission on Dams (WCD)*.

Maksimovic, C. and Tejada-Guibert, J.A. (Eds.) 2001. *Frontiers in Urban Water Management*, IWA Publishing.

Maksimov, C. Paris, 2001. *Urban Drainage in specific climates*. IHP-V Technical Documents in Hydrology, No. 40.

Hoyois, P. and D. Guha-Sapir. April 2003. *Three decades of floods in Europe: a preliminary analysis of EMDAT data.* Centre for Research on the Epidemiology of Disasters, Catholique University of Louvain, Working paper no. 197.

Gladwell, J.S. and Sim, L.K. 1993. *Tropical cities: managing their water.* IHP Humid tropics Programme Series no. 4, IHP-UNESCO.

Ashley, R.M. Building Knowledge for a Climate Change. 2003. *The impacts of climate change on the built environment.* Research Agenda. EPSRC.

MunichRe. 2001 Topics. Annual reviewer: *Natural catastrophes 200*, MunicRe, Munich.

2

Land Use and Urban Floods in Developing Countries

Carlos Tucci & Adolfo Villanueva
Institute of Hydraulic Research, Federal University of Rio Grande do Sul,
Porto Alegre, Brazil

ABSTRACT: Urban development in developing countries usually occurs with high population concentration in small areas, poor public transportation, lack of water facilities, polluted air and stormwater and floods impacts. Lack of integrated management of water supply, sewage, total solids, urban drainage and flood plain lies at origin of the problems.

The main types of floods which may impact urban areas are: Due to urbanisation: floods related to the increase of impermeable areas and man-made drainage facilities, such as conduits and channels; *Due to flood plain occupation*: when no reliable urban planning and regulation exists, population occupies the flood plain, usually after a sequence of low flood years. When a larger flood occurs, damage is very high and the municipality is requested to invest in flood protection for those areas.

Flood management is highly dependent on land use development. In many developing countries, the Urban Master Plan is mainly a compilation of tendencies of spontaneous occupation due to social and economics conditions. Understanding these aspects is fundamental to develop a sound flood management. These aspects are discussed, together with some Brazilian case studies.

2.1 URBAN DEVELOPMENT IN DEVELOPING COUNTRIES

In developing countries population growth is still high, and stabilisation is forecasted only after 2100 (The Economist, 1998, UN source). In the rank of the twenty most populated cities in the world 70% are in developing countries. Urban development in developing countries presents high population concentration in small areas, poor public transportation, lack of stormwater and sewage facilities, and polluted air and water. These poor environmental conditions are the main concern for life quality in these areas.

In most developed countries, domestic waste and quantitative aspects of urban drainage are no longer a major issue, the emphasis nowadays being on the control of urban drainage water quality. However, for developing countries access to basic sanitation is still a very important issue. Also, urban waste disposal without treatment is decreasing the amount of clean water available for supply, and new investments have to be made just to maintain supply and keep pace with population growth.

2.2 IMPACTS

Floods impacts on the population have been due to: i) population settles on risk areas such as flood plains and hill slopes during dry years and floods impacts them during wet years; ii) urban development increases impervious areas, which together with hydraulic improvements of urban creeks and minor drainage, increases the peak and volume of floods, overland flow and the frequency of impacts on the population. In January 2004, 87 people died in Brazil due to floods.

The main impacts of urban development on urban drainage and on the environment are: increase of flood peak, volume and frequency; degradation of urban areas due to erosion and sedimentation; water quality impact from wash-load of urban surface and solid waste.

2.2.1 Flood plain impacts

Flood plains are natural feature, found mainly in medium and large sizes rivers and problems appears when population settles on these areas. The main impacts on the population occur due to lack of both knowledge regarding the occurrence of flood levels and planning for space occupancy according to the risk of flood events.

A common scenario in an uncontrolled urbanisation is that population settles on the flood plains, during a sequence of years of low flood levels. When higher flood levels return, damage increases and the public administrations have to invest in population relief (see example in Box 2.1). Structural solutions such as dams, dikes, river channel changes have higher costs and are feasible only when damages costs are greater than works costs, or due to intangible social aspects. Non-structural measures have lower costs, but there are some political and administrative difficulties in their implementation.

2.2.2 Urban drainage floods

Urban drainage floods are those related to increase in impermeable areas, and man-made drainage such as conduits and channels. These floods are not natural events, they are man-created due to urban development. Usually land surface in small urban basins is a mixture of roofs, streets and others impervious surfaces. Runoff flows over these surfaces to the storm sewers at high velocity, increasing

Box 2.1. Floods at União da Vitória in Iguaçu River, Brazil

During 24 years (1959–1982) the floods levels at União da Vitória in Iguaçu river (see Figure 2.1) were below 5 years return periods (only one exception). It was the period of fast economical and population grow in Brazil. Floods after 1982 produced significant damages to the community.

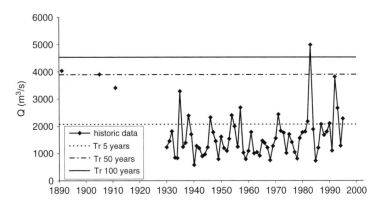

Figure 2.1. Maximum flood levels in Iguaçu river at União da Vitória (a basin of approximately 25,000 km^2, Tucci & Villanueva, 1997).

Table 2.1. Flood losses at União da Vitória
and Porto União (JICA, 1995).

Year	Losses US$ millions
1982	10.365
1983	78.121
1992	54.582
1993	25.933

peak flow, overland flow volume and decreasing groundwater flow and evapotranspiration. Under these conditions peak discharge increases together with flood frequency (Tucci, 2001 and Leopold, 1968; also Figure 2.2). Water quality deteriorates because during rain the surface is washed, increasing the pollution load in urban environment and to downstream rivers.

Cities have developed with little or no planning, and present Urban Master Plans usually do not consider the impact of urbanisation on drainage flow. Impervious areas and urban drainage flow are not regulated. City engineering departments commonly do not have the hydrologic capacities (know-how and/or staff) to

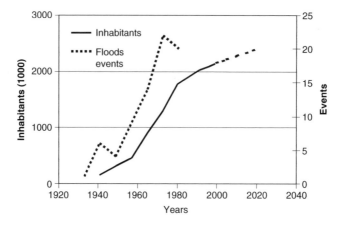

Figure 2.2. Population increase and flood events in Belo Horizonte, Brazil (Ramos, 1998).

cope with this problem through regulation, and engineering works – such as channels and pipes – are designed without taking potential downstream impacts into
account. It increases the flood frequency on the drainage with important economical losses and in some scenarios with loss of lives.

2.2.3   Institutional issues of flood management

The main issues on flood management are related to: occupation of risk areas on
the flood plains and on the hill slopes and the management of the flood frequency
increase in the urban drainage.

2.2.3.1   *Flood plain lack of management*
Risk areas are usually occupied by low income population, or by population with
higher income and with valuable facilities during a sequence of low annual peak
flows. Poor population usually invade this areas and during flood seasons receives
support from the public and when the city administration transfer this population
out of risk areas others take their place. This scenario usually does not occur when
the administration develops the area with public facilities.

The main issue is the lack of urban planning including flood risk areas management within the city development scenarios. In cities which have applied flood
plain zoning the main issue has been law enforcement on the development of public and private areas. On public areas there is the risk of invasion by the poor
income population while in private areas the issue is related to illegal development.

2.2.3.2   *Urban drainage management*
In developing countries there is a misconception in current urban drainage design and
management, because it is based on the concept of draining water from urban surfaces as quickly as possible, through pipe and channel networks; but this increases

peak flow and the cost of the drainage system. There is no control of peak increase at minor drainage level and most of these impacts will appear downstream in the major drainage. To cope with this problem, city and public administrations have to developed additional works such as larger channels in the major drainage and larger pipes in the secondary drainage network, and so on downstream. This type of solution has only transferred the flood problem from one section of the basin to another, with high costs. In addition, water quality decreases, since water carries a large, uncontrolled, amount of solids and is loaded with metals and other toxic components.

Since the 1970's in developed countries source control of urban drainage has been made by detention and retention ponds, permeable surfaces, infiltration trenches and others source control measures. It was implemented through county regulations and the cost of implementation is paid by the developer. In developing countries usually this type of control does not exist and the impacts are transferred downstream to the major drainage. The cost of the control of this impact is transferred from the individuals to the public, since the county has to invest in hydraulic works and structures to cope with downstream flood impacts.

2.2.3.3   *Main causes of flood problems*
The main causes of flood problems in developing countries are the following:
- urban development in developing countries cities occurs too fast and unpredictably. Usually the tendency of this development is from downstream to upstream which increases impacts and damages (Dunne, 1986);

- outskirts and risk areas (flood plains and hill side slope areas) are occupied by low income population without any infrastructure. Examples of spontaneous housing development in flood risk areas can be found in: Bangkok, Bombay, Guayaquil, Lagos, Monrovia, Port Moresby and Recife; hill sides prone to landslides: Caracas, Guatemala City, La Paz, Rio de Janeiro and Salvador (WHO, 1988);

- municipality and population usually do not have sufficient funds to supply the basics of water, sanitation and drainage needs;

- lack of appropriate garbage collection and disposal decreases water quality, and also the capacity of the urban drainage network due filling;

- lack of law enforcement or unrealistic regulations (see Box 2.2);

- lack of institutional organisation in urban drainage at a municipal level such as: regulation, capacity building and administration. In Asian cities there is a lack of (Ruiter, 1990): comprehensive project organisation and clear allocation of responsibilities; adequate urban land-use planning and enforcement; capability to cover all phases and aspects of technical and non-structural planning. Similar problems can be found in South and Central America.

Box 2.2. Urban occupation pressure on regulated areas

The city of Curitiba (Brazil) regulations prohibit land occupation of basins used for urban water supply and of flood prone areas, in order to protect them. Urban development has, to a certain extent, surrounded these areas thus increasing their real state value. Property owners are used to do the following: (i) clandestine development; (ii) encouraging poor population to invade their properties in order to break down the regulations and sell it to the municipality as a social solution. This usually occurs during election years when political pressure is higher.

   This situation is mainly due to lack of compensation for private landowners in the regulations, unfair since they have to pay land taxes without getting economic benefits from it. No taxes and some appropriate land use which does not degrade water quality would provide more incentive to land conservation.

2.2.3.4   Reliable flood management experience
Reliable experience in flood control in many countries has led to some basic principles in urban drainage management, which are:
– flood control evaluation should be done for the whole basin and not only in specific flow sections;
– urban drainage control scenarios should take into account future city developments;
– flood control measures should not transfer flood impact to downstream reaches, giving priority to source control measures;
– the impact caused by urban surface wash-off and others related to urban drainage water quality should be reduced;
– more emphasis should be given to non-structural measures for flood plain control, such as flood zoning, insurance and real time flood forecasting;
– management starts with the Urban Drainage Master Plan;
– public participation in urban drainage management should be increased;
– development of urban drainage should be based on a cost recovery rationale.

These principles have been applied in developed countries and are not fully used even in some of these countries. Urban drainage practices in most developing countries do not follow these principles.

2.3   INTEGRATED FLOOD MANAGEMENT

Urban drainage management is developed through an Urban Drainage and Flood Control Master Plan. Usually an Urban Plan is based on goals and objectives

Box 2.3. Tax incentive example

In Estrela (Rezende & Tucci, 1979) a study was prepared for the city together with the Urban Master Plan and included in the municipal regulations. After the legislation was implemented the risk areas were preserved and gradually the remaining population was removed to safe areas using taxes incentives. The tax incentives were the exchange of building construction area permits downtown for flood risk areas. Flood damages losses and population involved have decreased over the years since 1979.

related to well being of the population and environmental conservation. In urban drainage and flood plain management the main goals are:
– urban drainage planning by distributing flow volume allocation in time and space in the urban basin, based on urban space, hydraulic network and environmental conditions in order to reduce flood risks;
– control the occupation of flood plain areas through regulation and other non-structural measures;
– prevention and relief measures for low frequency floods;
– improve the urban drainage water quality.

Urban development conditioning factors are not discussed here since they belong to the Urban Master Plan (UMP) but there should be a strong interaction between this urban development document, drainage plan and other city plans. Land use and urban drainage are strongly related, and the UMP has to take into account the restrictions of the UDMP, as this one is a component of the former. Furthermore, there are other Plans related to it: water supply and sanitary control; garbage: street cleaning, garbage collection and its final disposal.

2.3.1   Urban drainage and flood management development

Non-structural measures: The non-structural measures are developed to control land occupation of the flood plains and to control the impact of urbanisation on the drainage.

Flood plain regulation usually restrict the use of these areas for new developments and plan new areas of occupation in the city using tax incentives (see Box 2.3).

Regulations related to urban drainage has the objective to control the downstream impact on peak discharge and to limit water quality degradation, taking into account social and economical conditions. The best regulation is that which increases public participation (see Box 2.4). Some of the basic aspects of this type of regulation are: the new development has to keep the peak discharge equal to or

Box 2.4. Public participation example

União da Vitória and Porto União are on the border of the States of Parana and Santa Catarina (Brazil) and are in fact one community (about 150,000 inhabitants) divided by an administrative divide. This urban area had been subject to frequent but low floods (see Box 2.1). In 1983 there was a major flood which had an important economical impact (sixty days of flooding). As in 1980 a large hydropower reservoir had been constructed downstream, the population began to blame the Electric Company (COPEL) which claimed that it was a natural flood and that the dam did not create any additional impact. But in 1992 another major flood took place, smaller than the 1983 flood but also with a high damage impact. It created a major conflict between the city and the Company. A NGO (Non-governmental organisation) was created by the population and the study was developed for this organisation whose goals were: diagnosis of the flood conditions, negotiations with the Company for operation rules and flood zoning planning for the city. The study brought some results and the negotiations improved the city's capability of dealing with floods. (Tucci & Villanueva, 1997).

below the pre-development scenario and limits for impervious surfaces in each development.

Structural measures: Urban drainage flood control is developed by sub-basins. In each sub-basin the existing flood impacts are evaluated for the risk and scenario selected. Based on these diagnosis, measures to control these impacts are designed. Usually the measures are a combination of detention ponds, conduits and channels, based on available space, existing drainage and topography.

2.3.2   Urban drainage plan outputs

The main outputs are:
– drainage control plan, including structural measures, and environmental and economical evaluations, for each sub-basin studied;
– regulation: non-structural measures included in the county legislation and/or in the city building code;
– urban drainage manual: the urban drainage manual is used to advise practitioners on city procedures and restrictions adopted by the city for urban drainage design;
– programs: these are long term actions which provide support to the Plan goals. The main programs usually are: hydrologic and water quality monitoring; public participation and capacity building at all levels.

2.4   URBAN DRAINAGE MASTER PLAN OF PORTO ALEGRE, BRAZIL

Porto Alegre is the capital of the State of Rio Grande do Sul in Brazil and has a Metropolitan area of about 3,7 millions inhabitants and the city county has about 2 millions inhabitants. The city is located by the Jacui delta, where the basin has about 80,000 km^2, and it is protected against floods by a system of dykes, with stormwater pumping stations designed and constructed around 1970. The city developed from downstream to upstream together with the peak floods. The present capacity of the drainage system is not enough to carry the upstream increase of flood peak and volume in some parts of the city.

Porto Alegre county covers an area of about 400 km^2 and comprises 26 basins. The Urban Drainage Master Plan was scheduled in steps. In the first part was developed the proposal of non-structural measures (legislation for new developments), review of the design capacity of the basins which are drained by pumping over the dyke system, and the Plan of six important basins of the city.

Non-structural measures: these measures were: (a) a new legislation on source control for developments which has been implemented since March 2000; (b) urban drainage education for urban drainage practitioners; (c) design manual. The county already had a department for urban drainage development and maintenance.

Structural measures: the Urban Drainage Plan for six basins was developed and below it is described some of the procedure developed for one of the basins.

Areia creek basin: The Areia creek basin has an area of about 12 km^2 and high population density (100 inhab./ha present and 200 inhab./ha planned). The lower sub-basins (L and I) in Figure 2.3 are drained by a pump station. The drainage from the upstream basin flows inside a pressure pipe (4 km long, which runs below the airport lanes and cannot be enlarged without huge costs) all the way to the Jacui Delta. The basin was divided in 11 sub-basins and the study scenarios were: present occupation and future occupation from the Urban Development Plan (based on population density).

Increasing pipe and channel capacity along major and secondary drainage would increase peak flow up to 140 m^3/s at a cost of US$15 millions (not taking into account downstream impacts). Using detention ponds in the major drainage at some public open spaces has a cost of US$8 millions and peak flow will be 42 m^3/s. Figure 2.4 shows the hydrographs for these scenarios in one of the sub-basins. In Figure 2.5 the detentions designed for this basin are presented.

This basin is a case that clearly shows the advantage of thinking basin as a whole, and not on point problems. There were several cases in which detentions were not possible or convenient, either for lack of space or because of high costs. Working with the whole basin, the solution was to increase drainage at those sub-basins, but that increased impact was compensated by enlarging detentions in other sub-basins, where there was space available, and costs were low. An example is sub-basin C, in which a detention was to expensive, and so the solution was

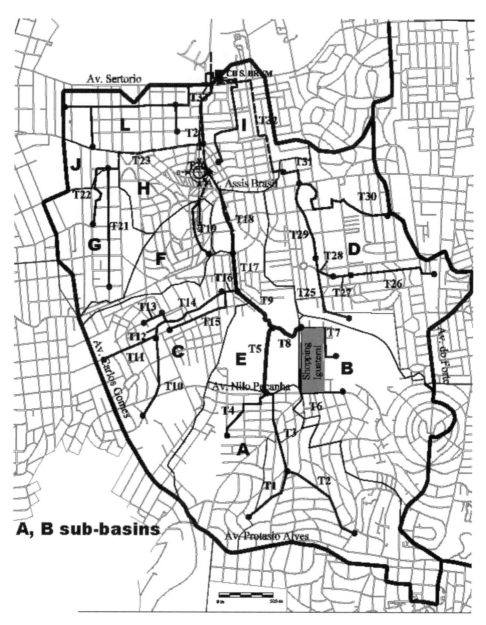

Figure 2.3. Areia creek basin.

larger pipes. To balance this, detentions in sub-basins A, B and E were enlarged, so as to keep discharge downstream more or less equal to the one with a detention in C.

Another clear advantage of having the UDMP for the basin is that now all drainage works in the basin fit into a general framework, which points both to

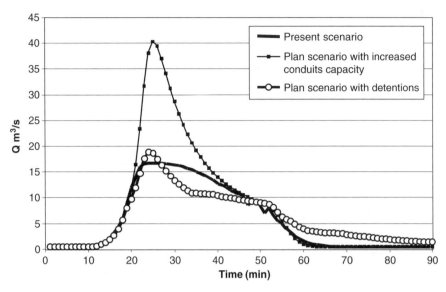

Figure 2.4. Hydrographs for the scenarios (future scenarios with 10 years rainstorm) basin A (IPH, 2001).

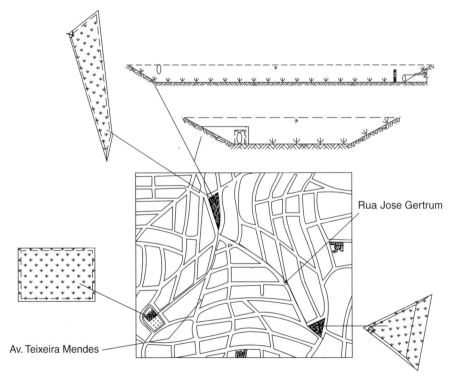

Figure 2.5. Detentions in sub-basin A (IPH, 2001).

local and general solutions; in other words, the solution of a problem does not create another one. As usually administrations do not have the money to execute all measures at the same time, being able to approach them individually, but within a coherent arrangement allows a more efficient use of available resources.

2.5 FLOOD MANAGEMENT OF METROPOLITAN AREA OF CURITIBA

Location: The Metropolitan Area of Curitiba (RMC) (State of Paraná, Brazil), has 2,5 million inhabitants. Most of this urban area was developed in the Upper Iguaçu River Basin which has a basin area of 1000 km² in J. Belem (Figure 2.6). The main tributaries are presented in Figure 2.6 and 2.7. Most of these tributaries have basin area of about 100 km². The highest urban concentration is in the Belem basin and other neighbours tributaries.

Floods Impacts: In RMC there are two types of floods:
– floods due to urbanisation which occur mainly in the tributaries of Iguaçu River such as Belem and Atuba. These floods usually occur in downtown Curitiba and in the highly urbanised parts of the cities in the Metropolitan area. In Figure 2.4 is shown the mean annual flood from a basin in Belém (72 m³/s, 42 km² and 40% of impermeable area) and its estimate of the pre-development mean annual flow (12 m³/s) from a regional extrapolation of non-urban basins. The flood increase due to urbanisation is six fold. In downtown Curitiba there were many channels transferring floods from one point to another resulting in a high cost for the city;

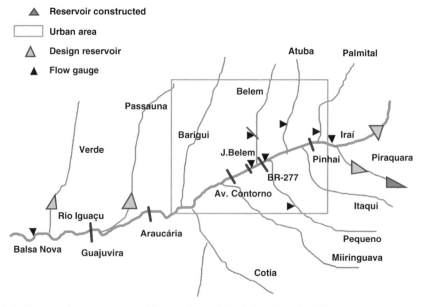

Figure 2.6. Iguaçu river at metropolitan region of Curitiba (Tucci, 1996).

- Iguaçu river has a large natural flood plain due to the small river conveyance and bottom slope. The main causes of floods in this area are: low river capacity, about 55 m³/s (return period 2 years) in BR-277 (Figure 2.6); flood plain occupation by the population; peak discharge increase due to urbanisation, mainly in Belém, Palmital and Atuba rivers; flow obstruction due to urban works such as bridges, land fill, poor drainage projects. During the flood months the hydrograph is damped by the storage capacity of the valley and the regional administration ruled against occupation of the floodplain. However, since 1980 there were heavy pressures for occupation of the flood plain. It was done through invasion of public green areas and by unapproved developments and occupation. In July 1983 and January 1995 two major floods occurred with severe damages of US$50.3 millions and 40.2 millions, respectively, (Consorcio, 1997). The 1995 flood had a seven day rainfall with more than a 100 year return period (largest in 110 years of data). The critical flood volume duration in the main river usually is seven days.

Flood Management: PROSAM was a Sanitary Program developed to cope with environmental impacts on the Metropolitan area of Curitiba and financed by World Bank. Flood Management was a component of this program which had the objective of developing flood measures taking into account urban constraints and minimising its economic and social impacts.

The planned project steps were the following:
- emergency measures: preliminary studies and actions which would minimise flood impacts. It was developed at the begin of the program and decreased some local floods;
- medium term flood control measures: which were the studies and actions in the Iguaçu river;

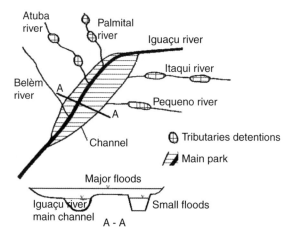

Figure 2.7. Flood management in Iguaçu river at metropolitan area of Curitiba (Tucci, 1996).

- long term measures: plan projects and actions in the tributaries together with the Urban Drainage Master Plan for the Metropolitan Area. The two first step were developed and the third step is almost concluded.

This text presents the lessons learned from the Flood Management in Iguaçu River. The usual approach would be to increase the Iguaçu river capacity to cope with the 50 or 100 year flood. Under these conditions the population would occupy the floodplain because of the flood frequency decrease just after the works. After a few years development of the upstream basin would change the hydrograph and increase the flood frequency and peak of the floods. In this future scenario there would not be more space for increase of the section width, since the flood plain would be occupied. In this scenario the flood control could be done only by dikes (with pump stations and internal drainage) or by deepening a river reach of at least 50 km, which represents a huge cost. This scenario occurred in the Metropolitan Region of São Paulo and the cost of deepening the channel was evaluated in more than a billion dollars in 1986!
    There are two main requirement in this problem:
- create a space for flow and storage;
- develop a process to control the population invading the flood plain (difficulties and lack of law enforcement).

The conceptual approach used was:
- in the main river (Iguaçu): The flood plain was preserved as storage area along the Metropolitan Area. A park 300 m to 1 km wide and total area of about 20 km^2 was planned and its implementation is under development. The boundaries of this area are a channel on one side and the Iguaçu main channel on the other side (Figure 2.5). The channel was planned in order to create a limit to urban settlement pressure and increase the river flow capacity for the Iguaçu basin in this area. The main recommendation was that the park has to be designed and implemented together with the channel construction. In Figure 2.6 is presented the Iguaçu main channel, small floods channel, linear park area and main tributaries.

- In the tributaries: Development of the Urban Drainage Master Plan for the Metropolitan Region using the following principles: (i) development of urban parks in the tributaries to damp the potential increase in the peak flow of the uncontrolled upstream area; (ii) implement counties regulation for the counties in the tributaries. This regulation should enforce source control in new developments. This strategic parks were planned to reserve storage areas in order to damp floods which could not be controlled by regulation and law enforcement. Tucci (2000) showed the requirement space for flood damping varies from 1 to 2% of the total basin area.

2.6   CONCLUSIONS

Flood and urban drainage management has to be developed within an integrated view with urban planning and other water facilities in order to cope with the main causes of the impacts and improve the quality of life of the population. This management in developing countries has many issues more related to institutional aspects than to technical ones (hydrologic, hydraulic or water quality). This issues are strongly related to population settlements, public participation and legal aspects, among others. In most of these countries institutional capacity is weak and without developing strong governance is impossible to have sound management.

This paper presents an overview of this matter in order to improve the awareness on the problem, aiming to advance in the development of sustainable solutions. It is also important to keep in mind that as each region or country has its own characteristics and peculiarities, there is not a unique solution. Experiences should be exchanged and adapted when suitable.

REFERENCES

Consórcio, 1997. *Parque e Controle de Cheias do Alto Uruguai: Obras Componentes do Sistema de Controle de Cheias*. Consórcio ENERCONSULT/ELC/TEI.

IPH, 2001. *Plano da bacia do Areia in: Plano Diretor de Drenagem Urbana de Porto Alegre. 1º Fase*. Instituto de Pesquisas Hidráulicas/UFRGS DEP/Prefeitura Municipal de Porto Alegre.

JICA, 1995. *The master study on utilisation of water resources in Parana State in the Federative Republic of Brazil*. Sectoral Report vol H – Flood Control.

Ramos, M.M.G. 1998. *Drenagem Urbana: Aspectos urbanísticos, legais e metodológicos em Belo Horizonte*. Master teases Engineer Faculty Federal University of de Minas Gerais.

Rezende, B.; Tucci, C.E.M. 1979. *Análise das Inundações em Estrela*. Technical report Estrela County. 30p.

Ruiter, W. 1990. *Watershed: Flood protection and drainage in Asian Cities*. Land & Water In'l 68: 17–19.

The economist, 1998. *Dirt poor*. The Economist Group. London UK.

Tucci, C.E.M.; Porto, R.L. 2000. *Storm hydrology and urban drainage*. In: Tucci, C. Humid Tropics Urban Drainage, capítulo 4. UNESCO.

Tucci, C.E.M.; Villanueva, A. 1998. *Controle de Inundações da cidade de União da Vitória*. Technical Report. CORPRERI. 135p.

Tucci, C.E.M. 1996. *Estudos Hidrológicos – Hidrodinâmicos do rio Iguaçu na RMC*. PROSAM SUCEAM Curitiba 2 volumes.

Tucci, C.E.M. 2000. *Coeficiente de escoamento e vazão máxima*. Revista Brasileira de Recursos Hídricos V5 n. 2.

WHO, 1988. *Urbanisation and its implications for Child Health: Potential for Action*. World Health Organisation. Geneva.

3

Cities, Lakes, and Floods. The Case of the Green Hyderabad Project, India

Joep Verhagen[1]
Disaster Mitigation Institute (DMI), New Delhi, India

ABSTRACT: Mitigating urban flood risks requires an integrated approach ranging from improved basin management to the construction of individual urban rainwater harvesting structures. This paper examines one of the potential measures to mitigate urban floods: *the restoration and conservation of urban lakes*. This paper starts with an overview of urbanisation in India and argues that the link between spatial planning and water has been broken as a result of uncontrolled urban growth and concurrent planning practices.

Mitigation of urban floods through the revitalisation of urban lakes is discussed on the basis of a short case study of Hyderabad (Andra Pradesh, India) where a large-scale lake conservation program has been initiated. The first results of the project are promising both in terms of urban floods and urban drought mitigation; indirectly the project contributes to urban poverty alleviation. The final section of this paper discusses challenges that need to be overcome and the potential for scaling up and replicating the Hyderabad project.

3.1   THE INCREASING INCIDENCE OF FLOODS

Floods have been recurrent phenomena in the region of South Asia albeit not only with negative impacts that are commonly assumed. People have chosen to live in flood plains and accept the floods risks as trade-off for the fertile deposits and, in some cases, easy access to good transport routes in the form of rivers. In China

[1] The author would like to thank the following persons for their time and sharing of their insights into the issue of cities, lakes, and floods: Mr. Robert-Jan Baken; Mr. Sanjay Barnela (Moving Images); Mr. Mihir Bhatt (Honorary Director, Disaster Mitigation Institute); Ms. Wilma van Esch (First Secretary Royal Netherlands Embassy), Mr. Henk Gijselhart (Team Leader Technical Assistance Green Hyderabad Environment program, Lake Component), and Mr. Subroto Talukdar (Senior Program Officer Royal Netherlands Embassy). The points of view and interpretations are attributable to its author alone.

for instance, 60% of the population live in flood affected areas which account for 80% of the countries revenues [Fox 2003].

However, over the last decades the frequency and the negative impacts of floods have increased sharply. Exceptional floods that affected China (1996, 1998), India, Nepal and Bangladesh (1998), are just a few examples. On average floods are more devastating than any other natural event and it is estimated that floods have caused globally a damage of more than US$ 250 billion [Loster nd]. Poor countries tend to be far more affected when hit by floods. For instance, the 2000 flooding in Mozambique cut the southern African country's GDP by 45 per cent, severe floods last year in Germany were blamed for just a one-per-cent decline in GDP [CNEWS 2003].

This upward trend can be largely attributed to socio-economic changes and the impacts of human activities on the environment. In Bihar[2] (India), for instance, the impacts of floods have increased due to the large-scale embankment of rivers. Recognition of the shortcomings of this traditional, engineering-based, approach of flood control has induced a gradual shift in favour of flood management. Flood management captures two important concepts: "the realisation that floods never can be fully controlled and that floods are part of the natural resources to be considered within the scope of integrated water resources management (IWRM)" [Fox, 2003: 11]. However, similar to IWRM the concept of flood management pays little attention to urban areas, despite rapid urbanisation, especially in the less-developed countries.

Table 3.1. Frequency and damage (in billions of US$ values 1998) of major floods globally during the last five decades.

	1950–59	1960–69	1970–79	1980–89	1990–99	Last 10/last 50
Number of major floods	7	7	9	20	34	4.9
Economic losses*	27.9	20.2	19.2	25.5	199.6	7.2
Insured losses*	–	0.2	0.4	1.4	7.4	–

* in billions of US$ values 1998.
Source: Loster not date.

---

[2] This example is derived from a video documentary by Moving Images titled "River Taming Mantras". This documentary illustrates that the large-scale embankment carried out by the Government of India (GoI) has had an adverse affect in three ways:
(a) Because of these embankments, fertile silt is no longer deposited on the flood plains leading to a degradation of the land.
(b) In case the embankment breaches, floods are hard difficult to predict and severe.
(c) As the level of the river water has become higher than the surrounding land due to deposition of silt in the riverbed, floodwaters cannot flow back into the river causing water logging.

3.1.1   Floods in India

Floods are the most recurrent disasters in India. Around 40 million hectares or 12% of the total surface area are under the threat of floods and even in the drought year 2002, 8.63 million hectares where affected by floods [GOI 2002]. Notwithstanding positive effects these floods have in some areas, floods cause a direct damage of US$ 240 million on average and in case of severe floods the damage can increase to over US$ 1.5 billion [UNEP, nd] equalling to 0.35% of the 1999 GDP [ADB 2001].

The area actually affected by floods have been steadily increasing during the last five decades. Most of these areas are found in the flood plains of the major and minor rivers, coastal zones, and low-lying urban areas. The five-year average of flood-affected area went up from 7.51 million hectares (1953–1957) to 14.85 million (1977–1982) [CSE 1991]. Floods mostly occur during or shortly after the monsoon period from July till September with an exception of coastal floods that are caused by cyclonic storms. During the monsoon, 75% of the annual rainfall occurs, often concentrated in a few days of intensive rain.

Although rural floods have been reported extensively, considerably less attention has been paid to urban floods. In recent years, urban floods occurred, amongst others, in Ahmedabad (1999, 2000, 2003), Mumbai (1999, 2000), Delhi (1999, 2000, 2003), Chennai (1999), Kolkata (1999), Bangalore (1999, 1999), Hyderabad (2000),

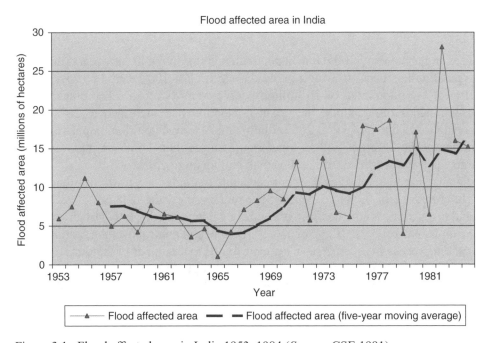

Figure 3.1. Flood affected area in India 1953–1984 (*Source*: CSE 1991).

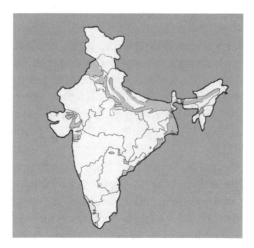

Figure 3.2. Flood plains in India.

and Guwahti (2003)[3]. Though the number of victims is limited, economic damage is huge. The damage of the Hyderabad and Ahmedabad floods is estimated to be around US$ 140 million [CSE 2000] and US$ 200 million respectively. However, the above list is far from complete as most floods in medium and small cities only get reported in the vernacular media.

3.2.   URBANISATION IN INDIA

Though the majority of the Indian population is still living in rural areas, the urban population has been growing rapidly during the last 100 years. In 1901, around 10% of the population or 25 million people were living in urban areas. At the beginning of this century, the urban population had increased to 28% of the population. This is equal to 285 million. A quarter of the urban population lives in one of the 27 cities with a population of more than 1 million, the remaining three quarters is spread over 3,600 small and medium cities.

Indian urban authorities have not been able to guide the growth of urban areas. '[In contrary], in the large share of the Indian cities, the delivery of residential land takes the form of squatting or semi-illegal land subdivision. Urban growth is an unplanned process.' [Baken 2000: 9]. This lack of control over the spatial development process in urban areas can be attributed to a number of factors:
- rapid growth of the urban population and high pressure on urban land have caused wide-spread speculation in land and a sharp increase of land prices.

---

[3] Sources: Deccan Herald, Asian Age, Hindu, BBC, The Telegraph, Catch Rainwater (Centre for Environment).

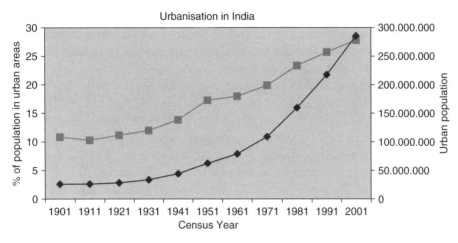

Figure 3.3. Urbanisation in India *(Source*: GOI 2001).

This, in combination with rampant political interference and vote bank politics, has made the commercial interests of (semi)-illegal land brokers the driving force behind spatial development rather than the urban master plans. The interests of a few often prevail over the larger common interest. Urban infrastructure is provided, under pressure of the new settlers, after the land has been occupied. Almost as an afterthought.

– urban master plans are basically land use plans that pay little attention to social and economic factors. Plans start from an assumed but unrealistic leading role of public agencies in the development of land and the supply thereof. They do not allocate the resources or specify responsibilities to realise them and hence are unrealistic from the onset [Baken 2000]. Moreover, long-drawn legal procedures make it almost impossible for urban authorities to obtain land needed to realise urban master plans.

– a large part of the legislation and regulations pertaining urban development and land-use stem from the pre-independence period and hence are ill suited for the current circumstances. Continuous amendments combined with the absence of a clear and overarching urban development policy have resulted in a legal labyrinth where few know their way around. Moreover, responsibilities of different public bodies are not clearly demarked and there is a lack of co-ordination between these bodies.

– for example, in Ahmedabad, the Ahmedabad Municipal Corporation (AMC) is responsible for the inner city, the urban periphery falls under the purview of Ahmedabad Urban Development Authority (AUDA) whilst urban lakes are controlled by the Collector of Ahmedabad district.

While there is still some degree of control over the spatial development process in metropolitan cities such as Delhi and Mumbai, the situation is worse in most small and medium cities. This uncontrolled urban sprawl has a number of important consequences with regards to urban floods.

Out of necessity, spatial development and water were closely interwoven in historic Indian cities. For instance, the historic cities of Varanasi and Delhi are located on safe spots along the river [CSE 1991]. Moreover, the urban watersheds were left intact and were often further developed with the objective to harvest rainwater and/or mitigate urban floods. The city of Jodhpur, for instance, has a complex water management system that has provided the city with sufficient drinking water for centuries. Jodhpur fort could store enough water to last for up to two years [CSE 1997].

The uncontrolled urban development of the last decennia has completely disrupted the existing urban watersheds. Natural drainage channels have been blocked, urban lakes filled up and encroached upon. The situation is further exacerbated by inadequate drainage systems and the poor maintenance thereof. Water logging in Delhi got aggravated because drains were not cleaned before the onset of the monsoon [Hindustan Times 2003]. Urban authorities often blame each other for this lack of maintenance. In Delhi, the Municipal Corporation Delhi (MCD), the New Delhi Municipal Corporation (NDMC), and the Public Works Department (PWD) kept on passing to buck to each other in 2003.

Finally, urban areas are characterised by a high area under impervious surfaces such as roads, pavements, houses, and so on. High rates of development along with the loss of soft landscape has led to high surface water sun-off rates. Moreover, there is a tendency in middle and high-class residential areas to pave roads whenever possible. This results in flash floods in the low-lying areas even after moderate precipitation.

The poor are disappropriately affected by urban floods. The high pressure on urban land has forced the poor to live in increasingly marginalised areas such a road sides, canal banks, low lying areas, and so on. Most of these areas are prone to flooding. Baken found: "Over time, squatments were pushed towards increasingly marginal areas: from plain land to hill slopes; from hill slopes to irrigation canal banks; to the margins of railway tracks, road and highway, and eventually to the beds and banks of the Krishna and Budameru Rivers, which are prone to serious yearly floods" [Baken 2000: 303].

Consultations with poor victims by DMI reveal the impacts of urban floods on the poor. Houses gets damaged, the incidence of water borne disease increases but maybe most importantly livelihoods get destroyed trapping the poor in an vicious cycle of ever increasing vulnerability. In some cases, the total financial loss equals the total annual income. Hence, it might take the poor months to overcome the impacts of even a short flooding period.

I live near a river, every time there is rain the house is damaged. Our scrap business also gets negatively affected as we have to stay at another place for 10–15 days and cannot work for a month. After every monsoon, we have to restart our business. Sometimes we can't even feed our children during that period.

Gauriben Revabhai Senma, Ahmedabad

Our house is located near the river. In 2002, our house was washed away and we lost our entire stock of chocolate, *pan masala* (chewing tobacco) that we used to sell in a small shop attached to our house. The floods caused serious economic problems for us. There was not enough food and we had to keep our children at my sister-in-laws house and stopped sending them to school.

I live near the riverbed. We, my husband, my 6 yr. old daughter, and I, survive by making and selling food. During the 2000-floods, food worth Rs 5,000/- (approximately Euro 100) by the water. I lost my daughter's birth certificate and therefore she was denied admission in school. All my family members contracted cholera. Now we live in constant fear of the water.

3.3 THE GREEN HYDERABAD PROGRAM

3.3.1 Program background

Hyderabad is located in the South of India and is the capital of the State of Andra Pradesh. In the recent past, the city has emerged as one of the main centres of the Indian software industry. During the last 40 years, the population of Hyderabad has grown rapidly from 1.4 million (1961) to almost 5 million at present. Simultaneously, population density increased by 50% and consequently, land use changed dramatically in large parts of the city.

Hyderabad receives an average rainfall of 750 mm per annum, mostly from the Southwest monsoon in the period of June to October. The physiography is dominated by bare hillocks along with plain lands. Ground elevation varies from 610 to 487 meters. The topographic setting combined with limited vegetation in the area, results in a large and rapid runoff during rain.

Historically, urban lakes have been used as a buffer for runoff and as storage of rainwater for later use. In the past, Hyderabad counted more than 500 artificial and natural lakes. Artificial lakes were formed by constructing bunds across seasonal streams. Lakes were interconnected so that during high intensity rainfall, higher lakes overflowed in lakes lower down the watershed. Stored rainwater was used for domestic and irrigation purposes.

As a result of the rapid and uncontrolled urban growth and the consequent changes of land use, the functioning of this network of urban lakes has deteriorated rapidly: lakes have been encroached upon and polluted, and linkages have been blocked. The number of water bodies, once 530 is down to 150 at present [CSE 2000]. One of Hyderabad's most prominent lakes, Hussain Sagar Lake, has shrunk from 75 hectares to 25 hectares and lakes such Mir Jumala Tank, Ma Saab Tank, Batkamma Kunta are lost forever.

The combined effect of increased impervious surface and the deteriorating quality of the urban watershed has resulted in and increased incidence of urban floods. Rainfall is the predominant cause for urban floods. During the last 100 years, Hyderabad has been hit by major floods in 1908, 1915, 1916, 1933, 1962, 1970, and 2000. However, urban flooding is no longer restricted to the years with heavy rainfall. Rainfall exceeding 5 cms in 24 hours is enough to clog up drains, fill up streets, cause water logging, and disrupt life. Intensity over 15 cms causes large-scale floods. At the same time, Hyderabad is facing severe water shortage during the summer period.

Thus there is strong need to improve the management of the urban watershed in Hyderabad. Restoring urban lakes and re-establishing their links is part of the effort to mitigate urban droughts as well as urban floods.

3.3.2   The Green Hyderabad Project[4]

The first phase of the Green Hyderabad, implemented by the Hyderabad Urban Development Authority (HUDA) with support from the Royal Netherlands Embassy (RNE), mainly focused on the upgrading of the urban environment through large-scale tree plantations. Women groups raised saplings and large-scale tree plantation drives were organised with the participation of school children, community groups, and the private sector.

The present project, initiated in 2001, takes a broader perspective. It seeks to restore and conserve 85 urban lakes with a total surface of 1,300 hectares. Secondly, the project aims to green more than 12 thousand hectares of urban area. Though both objectives are interlinked the remaining part of this paper focus on the first objective of treating and conserving urban lakes.

Urban lakes were divided into tentative categories. Category I lakes are heavily polluted and need extensive treatment. Sewerage treatment plants will be used to treat raw sewage that is feeding these lakes. In total, 21 of these Category I lakes have been identified. The remaining 64 lakes are less polluted and water will be treated through constructed wetlands.

Simultaneously, lake bunds are widened and strengthened; inlet and outlets identified, sluices and surplus weirs are improved. On the shores of the lakes,

---

[4] The case study is based on discussions and documentation provided by project representatives and of the Royal Netherlands Embassy.

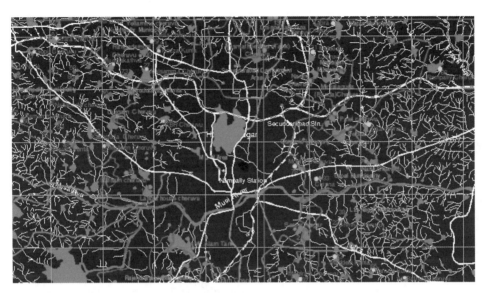

Figure 3.4. Hydrological map of Hyderabad.

greenery or recreational facilities are planned to ensure that lakes are not encroached upon in future. For a number of Category II lakes, formal contracts have been closed with women groups for the use and management of the lakes. During the summer period when most lakes are dry, women have the right to use these lakes for vegetable growing and so on. At an institutional level, the capacity of HUDA, and the involved HUDA staff, is further strengthened in fields such the use of Geographic Information Systems (GIS).

The project is expected to have positive impacts for the urban poor. Around 30% of the project lakes are situated in poor neighbourhoods and will contribute to the mitigation of floods in these areas. Hence, the project will result in a decrease of the damage to the habitat and livelihoods of the urban poor, improve their health situation, help them escape from the vicious circle of increasing vulnerability and decreasing coping capacity.

3.3.3   Challenges

The social and political environment prevalent in Hyderabad posed and poses a number of challenges for the successful implementation of the program. Most of these challenges are applicable to most Indian cities.

3.3.4   Political will and supportive legislation

Political will is essential for the success of the program. Support was guaranteed from the inception phase of the project as the improvement of the urban environment in Hyderabad always enjoyed the strong support from the Chief Minister of

Andra Pradesh, Chandrababu Naidu. Moreover, citizens have been filling 'public interest petitions' to enforce protection of the urban lakes [CSE 2000b] thereby generating the necessary political pressure to ensure the aforementioned support.

Secondly, urban forestry is protected by a special act that bans the cutting of trees in urban areas in Hyderabad. However, there is no legislation in Andra Pradesh, and the rest of India, to protect urban water bodies. Moreover, conservation of urban lakes did not get any mention in the latest National Water Policy. In a number of Indian cities, urban lake protection has been legally enforced by concerned citizens who were affected by floods or a deteriorating urban environment. In most cases, this legal action has been taken by people belonging to the urban middle and upper class.

3.3.5   Institutional arrangements

As argued above, institutional arrangements and jurisdictions are unclear in most Indian cities; different agencies are responsible for different parts of the urban watershed. The communication and collaboration between these agencies is often poor.

The Hyderabad Urban Development Authority (HUDA) acts as the nodal agency for the implementation of the Green Hyderabad Project. For this purpose, the authority over the urban lakes was transferred from the Irrigation Department to HUDA. A high level steering committee is guiding the project whilst an Executive Committee has been created to ensure efficient co-ordination at the ground level.

However, a well-maintained storm water drainage system is an essential part of the urban watershed. In large parts of Hyderabad, this system is outdated and badly maintained. As this falls under the authority of the Municipal Corporation of Hyderabad (MCH), HUDA is not in a position to take the necessary action directly. Secondly, after completion of the project HUDA needs to transfer the lakes to the local authorities, which might not be fully prepared for the task of managing these lakes.

3.3.6   Interlinking of lakes

Lakes will be interlinked with underground pipes. Though this solution is more expensive it was preferred over open connection as it was felt that open connections would get blocked easy and hence be rendered ineffective in short period of time.

3.3.7   Protection and encroachment of the lakes

Treated lakes are protected by a strengthened bund and a zone of greenery. So far this has been sufficient to halt the encroachment of lakes, however, it also stops part of the livelihood activities such as livestock rearing, washing of clothes, and so on. The poor are most affected by this.

However, in general, the encroachment of existing lakes remains the foremost obstacle for the treatment of lakes and it seems unrealistic to assume that encroached lakes ever will be returned to their original state.

3.3.8   Finance

It is not possible to finance lake conservation through cost recovery mechanism, as it is difficult to levy a user fee to people who benefit from upgraded lakes although the benefits are considerable. Hence, urban lake conservation projects depend on external financial support. In case of the Green Hyderabad project, a major part of the costs are born by the RNE.

No long-term arrangements have been made yet for the maintenance costs of the lakes. An additional charge on land or house tax will have limited returns, as only a small part of the population pays these taxes. Levying house tax is seen a *de facto* legalisation of an often (semi)-illegal land occupation, hence urban authorities are often reluctant to impose these taxes.

Public private partnership might provide an opportunity to finance the maintenance of the lakes. Firstly, the shores of the lakes could be leased to entrepreneurs who could exploit the surroundings of the lakes for recreational purposes. Secondly, property prices around the lakes have risen considerably after the treatment of lakes. Hence, Resident Welfare Associations (RWA) around the lakes have a strong stake in the maintenance of the lakes and can be asked to make financial contributions.

3.4   CONCLUSION AND RECOMMENDATIONS

Urban planning practices, or more accurately the total lack thereof, have had a detrimental impact on the quality of urban watersheds in Indian cities. Land brokers are the driving force behind urban spatial development rather than the urban master plans. The interests of a few often prevail over the larger common interest. Urban infrastructure is provided, under public pressure, almost as an afterthought. As a result, the existing link between spatial planning and water has been broken and, consequently, the incidence of urban floods and urban droughts has increased. The urban poor are hit hardest by this.

The Hyderabad case study confirms the potential of conserving urban lakes as a means to mitigate urban floods. The poor, and women in particular, are likely to benefit from the Green Hyderabad Project. However, impacts (negative and positive) need to be further quantified and qualified. For instance, it is not clear yet to what degree urban floods will be mitigated, how many of the urban poor are likely to benefit, or to what extent groundwater tables can be recharged.

Urban lakes cannot be treated in isolation but need to be considered as an integral part of urban watersheds. Well-functioning storm water drainage is necessary and linkages between lakes are essential parts of this watershed. Moreover, urban areas need to become part of basin management plans and, at a micro-scale, individual rainwater harvesting has to be promoted.

Political support is a pre-condition for any effort to replicate or up-scale lake conservation programs. The urban middle class has the potential of mobilising this support. Secondly, effective legislation to protect urban lakes and adequate institutional arrangements are necessary.

External finance is needed for the actual implementation of lake conservation programs. However, resources for maintenance can be generated through Public Private Partnerships. Moreover, costs can be reduced by focussing on 'low-cost high-impact' interventions by focussing on the fringe of the city or in small and medium-scale city where lakes are relatively unspoilt and less polluted.

Finally, little data and information on urban floods and the mitigation thereof is available. For instance, in India, no comprehensive data on urban floods is available. Documentation of direct and indirect impacts of urban floods, and best – and worst – mitigation practices is needed to put this growing problem on the agenda of policy makers and to generate a body of knowledge to assist practitioners.

Glossary

ADB	Asian Development Bank
GDP	Gross Domestic Product
GIS	Geographic Information Systems
GOI	Government of India
HMC	Hyderabad Municipal Corporation
HUDA	Hyderabad Urban Development Authority
IWRM	Integrated Water Resource Management
RNE	Royal Netherlands Embassy
RWA	Resident Welfare Association

REFERENCES

Agarwal Anil and Narrain Sunita (eds) 1987. The Third Citizens Report: *Floods, Flood Plains and Environmental Myths*. Centre for Science and Environment. New Delhi, India.

Agarwal Anil and Narrain Sunita (eds) 1997. The Fourth Citizens Report: *Dying Wisdom. Raise, fall and potential of India's traditional water harvesting system*. Centre for Science and Environment. New Delhi, India.

Baken, Robert. Jan 2000. *Plotting, squatting, public purpose and politics. Land market development, low-income housing and public intervention in Vijayawada and Visakhapatnam, India* (1900–1993). PhD thesis. Free University of Amsterdam, The Netherlands.

Centre for Science and Environment 1991. *Floods, Flood Plains and Environmental Myths*. State of India's Environment. A citizens' Report 3'. CSE, New Delhi, India.

Centre for Science and Environment 2000a. *Catch Rainwater: The CSE campaign for People's Water Management.* Vol. 2. No. 5, 2000. www.rainwaterharvesting.org Centre for Science and Environment. India.

Centre for Science and Environment 2000b. Hyderabad; Analysis. In *Down to Earth* June 30, 2000. Centre for Science and Environment. New Delhi, India.

CNEWS, 2003. *Global Warming-induced Weather Expected to Get Worse.* http://cnews. canoe.ca/CNEWS/Science/2003/02/27/33582-ap.html

Fox Ian B. 2003. *Reducing the Vulnerability of the Poor to the Negative Impacts of Floods.* Working Document. ADB. Manila, Philippines.

Government of India 2001. Census 2001.

Government of India 2002. Flood Situation Report 2002. India Metrological Report. Government of India.

Hindustan Times. *People in a jam, MCD, policy play blame game.* Hindustan Times 11-07-2003. India.

Loster, Thomas, not dated. *Flood Trends and Global Change.* Geoscience Research Group Munich Reinsurance Company.

UNEP no dated. Asia-Pacific Environmental Outlook. United Nations Environment Program. www.rrcap.unep.org.

4

Co-operation within Europe on Flood Management and Spatial Planning

Roelof Moll
Royal Haskoning, Rotterdam, The Netherlands

ABSTRACT: In densely populated low-lying areas floods have always been a major concern. In the Middle Ages in the Netherlands, Water Boards were founded, primarily for flood control. Modern organisation of water management has been developed from these roots. Recent floods in Europe and expected climate change have brought the flooding issue more prominent on the Agenda. Two important insights have been grown and recognised in the last decade:
1. spatial planning rather than traditional dike enforcement offers sustainable solutions to the flooding problem;
2. flood management should be carried out on a river basin scale.

Water management on a river basin scale has been formulated as one of the kernels of the European Water Framework Directive. This WFD forms the actual focal point of far-reaching on-going re-organisation of water management all over Europe. The acknowledgement of the importance of spatial planning in relation to flood management has been the core of the EU-programme IRMA (International Rhine Meuse Action Programme), which was implemented in the period 1988–2002.

This paper highlights the role of the IRMA programme in the development and implementation of concepts for flood management in Western Europe. Whereas IRMA had all its activities in the river basins of Rhine and Meuse, the results of the programme and the lessons learned are expected to have an impact on flood management initiatives in other basins within Europe.

4.1 THE IRMA PROGRAMME

The Rhine and Meuse riparian EU countries intending to intensify co-operation on flood management through spatial planning have together written a so-called Joint Operational Programme (JOP) IRMA.

Main objective of this IRMA programme has been formulated as:

To prevent damage caused by floods for all living creatures and important functions in the catchment area of rivers and therefore create a spatial balance between the activities of the population in the areas, the socio-economic development and sustainable management of natural resources.

This objective combines three elements: spatial planning, water management and damage prevention.

This main objective has been elaborated along three principles:

1. retain water: water should be retained in the catchment areas of the rivers as much as possible;
2. give space to water: the rivers should have space to discharge at the lowest possible risk level and high-water should have the opportunity to flow into retention and flooding areas;
3. raise the awareness for high waters: knowledge has to be improved, models developed, studies undertaken, legislation drafted and the most favourable conditions have to be created.

These principles have been used to define subsequently three project Themes:

Theme 1: measures within the catchment area;
Theme 2: measures within the major riverbeds of Rhine and Meuse;
Theme 3: improve knowledge, public awareness and (trans)national co-operation.

Under Theme 1, the following measures have been included:

1.1 Restoration of natural courses of tributaries and their overflow areas;
1.2 Indirect discharge of rainwater from residential and industrial areas;
1.3 Creating retention and overflow areas.

Under Theme 2, the following measures have been included:

2.1 Preservation, restoration and measures intended to use water retention areas and enlarge wetland retention areas.

Under Theme 3, the following measures have been included:

3.1 Development of models and development of spatial planning;
3.2 Promotion and implementation of good practices of monitoring, early warning and protection systems;
3.3 Identification of sensitive areas, raising awareness and increasing know-how.

The European Commission has approved the JOP of IRMA in December 1997. This implied that an amount of 140 million Euro had been approved and allocated for funding of projects under IRMA by the EU. This amount represents a maximum of 37% of the total project cost. Total investments under IRMA exceed therefore 400 million Euro.

4.2   EXAMPLE PROJECTS

4.2.1   Theme 1: Riparia

Large sections of the middle reaches of the river Rur in Germany were canalised
and shortened in the 20th century to create more space for farmland. The current
began to flow more rapidly as a result. In high water periods, the river thunders into
the Netherlands at Roermond, increasing the risk of flooding in the Meuse valley.
The project intervention consisted of a set of nine measures, including: the recovery
of flood plains, restoration of the original river course, connecting old river arms
and dredging, re-forestation, relocating dams and weirs, nature development and

Figure 4.1. IRMA Project map Bakenhof project.

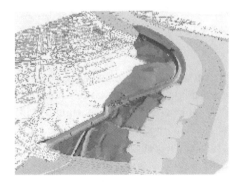

Figure 4.2. Old situation.   Figure 4.3. With dike relocation.

Figure 4.4. View on newly constructed side-channel.

dike repair. More than 1.9 million m^3 retention capacity has been created. The project may be considered a highly successful mixture of transnational flood protection and regional nature conservation. A co-operative agreement with the agricultural organisations played a key role in the success of the project.

4.2.2   Theme 2: Bakenhof

One of the bottlenecks in the Dutch branches of the Rhine that impedes the discharge of water is a narrow stretch of the Lower Rhine east of the city of Arnhem. At the Bakenhof location, in a river bend, the river has very little space. The project intervention consisted of a 200-m. inland relocation of the river dike, including a redistribution of land-use functions. In the restored flood plain, nature development has been stimulated, a small side-channel and footpath have been created. Before removing the old dike, archeological research has been carried out, producing interesting insights in the earlier use of the site. Fortunately, the volume of contaminated soil was below expectations. The local effect on design high water levels amounted to 7 cm. Through implementation of a participative approach full co-operation of stakeholders was achieved, leading to a timely project execution.

4.2.3   Theme 2: Oosterbeek

An important bottleneck in the discharge function of the Lower Rhine forms the earthen railway embankment near Oosterbeek. This embankment causes backwater effects extending 20 km upstream into the city of Arnhem. The embankment dissects the relatively high-lying floodplain. The project intervention consisted of the construction of a permeable railway bridge and the excavation of the floodplain. Specific problems in this location are the noise pollution in this densely populated and the considerable amount of explosives still lying in the ground since the Second World War. The intervention is rather costly, and forms in fact the most expensive IRMA project. The start of the works has been delayed due to the problems mentioned. To date, the works are well under way, and are expected to contribute to a systematic lowering of high water levels with no less than 15 cm locally.

4.2.4   Theme 3: Sponge

The SPONGE umbrella programme was set up to develop tools that planning and water management authorities can use to take measured decisions with respect to flood protection. The aim has been formulated to provide a single platform for more than 50 researchers and policy-makers working for more than 30 different organisations in the Rhine and Meuse states. A cluster of thirteen interrelated projects has been set up, emphasising climate change, water management, and ecology and policy recommendations.

Through the project intervention it has been demonstrated that the most effective strategy for preventing flood damage and raising public awareness of flood risks is to incorporate flood protection in spatial planning targets. There are various ways of pursuing this strategy. At times, difficult decisions that affect private and other local interests will be unavoidable.

4.3   LESSONS LEARNED

The IRMA programme has brought to the participants valuable experience in practical transnational co-operation in water management. IRMA typically has been designed as an implementation oriented programme, not as a study programme. Theoretical statements on the benefits of international co-operation and of working on a river basin scale have obtained a practical value within the context of IRMA activities.

The fact that the EU was funding on the average 37% of the project cost within the programme proved to have a large impact on the development of the contents of project proposals. Following the principle of subsidiarity of EU funding, which implies that the EU is not aiming to take over funding of projects having a strictly national character, a European or transnational dimension was required for each proposed project.

Which practical forms can transnationality take? Transnationality could be the geography of the project location. This simple form of transnationality occurred when the project area covered for instance a border crossing of a river. Transnationality, however, could also take other forms. An example of a completely different form was the inclusion of a set of joint workshops of different national sub-projects within one IRMA-application. The common river basin usually was the common denominator for such a transnational project. This type of projects led to many useful co-operation forms on a working level, supporting the development of formal co-operation mechanisms. Through project workshops, an intensive exchange of experience took place. Thus potential project applicants were strongly stimulated to develop the transnational dimension of their thinking. This undoubtedly led to a broader use of spatial planning principles for flood management in the IRMA countries. Positive and negative project implementation experiences were thus quickly shared between the IRMA participants.

A specific example of a transnational dimension is the project Bislicher Insel. In this project, 6 km of dike relocation and construction of a flood storage area of 50 million m^3 has been realised along the Rhine in Germany, using EU funds allocated for Germany and for The Netherlands. The contribution of this project to flood safety in the Netherlands was thus acknowledged.

Applicants for IRMA subsidy had to be Governmental Agencies. This brought the Agencies in a new role. They had to set up a project and financial administration allowing transparent justification of expenditures of EU money. Inexperience on this led to delays in project implementation.

Other delays occurred. Spatial planning procedures often take a long period, especially when stakeholders make use of their legal possibilities to protest. In some areas private land ownership proved to be an obstacle blocking timely project implementation. A very practical cause of delay in the Arnhem-Oosterbeek area proved to be the remnants of the Second World War: non-exploded ammunition was still abundantly spread out under the earth in the whole area.

The fact that the EU subsidy was available for a limited time period only definitely has accelerated the implementation of many project ideas. Only Right Now was the opportunity! Although the EU subsidy only covered a part of each total project budget required, the effect on speed of implementation can be clearly acknowledged. This also has led to a quick and widespread adoption of the spatial planning approach to flood management.

One of the EU requirements for project funding is sufficient exposure. The EU has even issued a brochure prescribing minimum sizes of plaques and notices to be placed on a project site! The exposure of projects, however, served a purpose, as the concepts of spatial planning solutions to flood problems have thus been actively promoted, supporting public awareness.

4.4   NEW CHALLENGES

IRMA has issued a publication "Water After IRMA" describing the possibilities for obtaining EU-funding for spatial planning oriented water management projects after the IRMA programme period in the Rhine and Meuse river basins. Relevant EU-programmes include: Interreg IIIA, IIIB, IIIC, LIFE, LEADER, RDP, 5th Framework Programme and others.

Whereas the EU may be seen as a catalyst to fund projects, it may be clear from the results of the IRMA programme, that the added value of EU involvement is far broader than that. The acceleration in exchange of new spatial planning concepts and experience in project implementation cannot be underestimated. The impact of the EU on water management and water legislation within Europe is underlined by the wide introduction of the EU Water Framework Directive. This Directive integrates many older EU directives, but still falls somewhat short on particular the flooding issue. It may be expected that the recent "Initiative on Floods" (Best Practice on Flood Prevention, Protection and Mitigation) taken by water directors of several European countries will result in an extension of the EU WFD or on additional guidelines on flood management related to spatial planning.

REFERENCES

Directive 2000/60/EC of 23 October 2000. 22.12.2000. *Official Journal of the EC.*
Draft. 2003. *Best Practices on Flood Prevention, Protection and Mitigation.*
IRMA Secretariat, Ministry of VROM, The Hague. 2002. *Water After IRMA.*
IRMA Secretariat, Ministry of VROM, The Hague. 2003. *IRMA Best Practices.*
Ministry of Transport, Public Works and Water Management, The Hague.

5

Spatial Measures and Instruments for Flood Risk Reduction in Selected EU Countries – A Quick Scan –

Willem Oosterberg
Ministry of Transport, Public Works and Water Management, Directorate-General of Public Waters and Water Management. Institute for Inland Water Management and Waste Water Treatment RIZA, Lelystad, The Netherlands

Jasper Fiselier
DHV Group, Amersfoort, The Netherlands

ABSTRACT: Traditionally, flood risk reduction in the Netherlands relies heavily on technical measures such as dykes and pumps. Due to recent instances of high river levels, excess rainfall and the prospect of climate change the Dutch flood prevention policy now places more emphasis on spatial measures. Finding "space for water" is not an easy matter for a densely inhabited country like The Netherlands; therefore there is keen interest to learn from the experience of other countries. To this aim, a quick scan has been performed on the experiences in this area in Austria, Belgium, England, France, Germany, Hungary, Italy, Poland and Sweden.

The quick scan focusses on spatial measures to reduce flood hazard, measures to reduce the flood damage potential, and on the juridical and economic instruments in support of these measures; measures and instruments for emergency management were not considered. Information was retrieved via internet and a questionnaire.

Nearly all countries that were included in the quick scan have faced major floods in the last decade. This has put flood risk reduction high on the political agendas.

Technical measures still form the backbone of flood prevention in all selected countries, but in most countries there is a strong interest in and growing experience with the application of spatial measures to reduce flood hazard and in the reduction of flood damage potential.

In most countries there are definite plans for the realisation of spatial measures, but actual realisation is still only small scale. Retention measures in the catchment are common in Germany, Italy, Austria and France. Examples of the application of spatial measures in polders or low lying areas were not found outside The Netherlands. Measures for infiltration in urban areas are well developed in Germany. Examples of retention areas and protected floodplains were found along rivers in Germany, Italy, Belgium and Hungary. Generally, these areas are situated in sparsely populated regions. Creating new retention areas and regaining floodplains generally requires tailor-made designs that attract sufficient local support and outside funding.

Restrictive hazard zoning in combination with building regulations are important instruments for limiting the flood damage potential in France, Italy, Austria and Germany. In England and Belgium, central authorities have obligated local authorities to take flood risks seriously in their land use decisions. Instances of private contracts to accept floods have been found in France and Germany. Germany is attempting to decrease urban runoff by imposing a tax on impermeable surface in urban areas.

The way in which flood damage compensation is arranged differs greatly from country to country. Some countries have a state compensation scheme; others rely entirely on private insurance; many countries have a mixed system. The high costs incurred in recent floods, that had to be covered by national authorities and the EU, has led to an increasing interest in private insurance.

5.1   INTRODUCTION

Floods appear to be becoming a more acute risk in many countries in Europe. This is illustrated by recent floods of large rivers, such as the Rhine, Elbe, Oder, Tisza and Danube and more local flood events along rivers in England and the Mediterranean. In The Netherlands, recent instances of high levels of the rivers Rhine and Meuse and events of excess rainfall, as well as the prospect of climate change, have led to a change in the water policy. This new policy now aims to find "space for water", rather than relying on technical measures such as higher dykes and more pumps.

Claiming space for water in a densely inhabited country like The Netherlands is not an easy matter. It requires the careful application of spatial planning procedures and juridical, economic and communicative instruments, and it requires a shift in the traditional view that flood problems can be kept under control by technical measures. There is keen interest to learn from the experiences in other countries on topics such as:
– examples of claiming and maintaining space for water;
– understanding what instruments were used and what factors determined the success of these instruments;
– making contact with persons and institutes who are working in this field.

In this paper we give the results of a quick scan that was performed by RIZA and DHV on the experience in the following countries:
Austria, Belgium, England, France, Germany, Hungary, Italy, Poland and Sweden.

5.2   QUICK SCAN

5.2.1   Scope of the quick scan

We have used the following definitions:
- Flood hazard: the probability that a certain area will be flooded.
- Flood damage potential: the sum of possibly damaged assets in the area at risk.
- Flood risk = flood hazard * flood damage potential (which are designed to influence human behaviour).

The scope of the quick scan is the application of spatial measures to reduce flood hazard and measures to reduce flood damage potential, and the policy instruments in support of these measures. Measures and instruments for flood preparedness (flood forecasting and emergency response) have not been considered. The quick scan is descriptive and not prescriptive – it asks the question of how measures and instruments are applied in practice, and not how they should be applied.

In the framework of IRMA-SPONGE valuable in-depth studies have been performed on spatial planning and flood prevention in the Rhine-bordering countries Switzerland, France, Germany and The Netherlands[5]. The study evaluates the inclusion of claims for "space for water" in regional and local spatial plans in these countries as well as the use of supporting (soft) instruments such as transboundary co-operation, economic instruments and information management. A conceptual framework is presented for the interplay of spatial planning and water management, which is used to specify further the scope of the quick scan (Figure 5.1).

5.2.2   Choice of countries

Countries were selected on the basis of the following considerations:
- a focus on neighbouring countries with whom we share transboundary rivers (Belgium, France, Germany)
- a focus on European countries because of a similar economic development, a similar climate and shared EU initiatives related to flood management
- inclusion of some new member states (Hungary, Poland)
- available funding did not allow coverage of all European countries.

5.2.3   Retrieval of information

A questionnaire was collated (see Table 5.1) which was put to contact persons – via e-mail and telephone – and formed the basis for searches on internet.

---

[5] IRMA-SPONGE 2001 project 5: "Spatial planning and supporting instruments for preventive flood management" The report can be found at http://www.irma-sponge.org/publications.php.

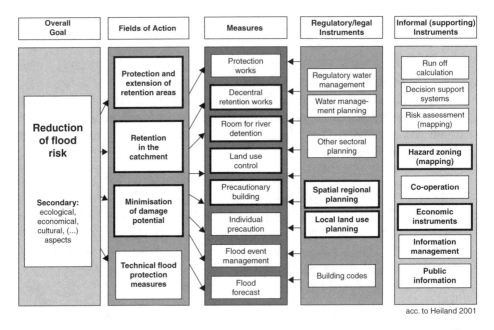

Figure 5.1. Goals, measures and instruments for preventive flood management (from IRMA-SPONGE 2001) and aspects included in the quick scan (bold figures).

5.2.4   Presentation of information

As the information was provided by various people and the amount of information on internet varies per country, omissions and differences in emphasis are inevitable. Opportunities for addition and verification were very restricted. As a consequence the information presented does not give an in depth and complete picture of the way in which spatial planning contributes to flood management in the investigated countries, but in the authors opinion it does offer a wide range of relevant solutions and instruments adopted in the EU.

5.3   RESULTS

5.3.1   Types of floods

Table 5.2 gives an overview of the main flood types experienced in the 10 countries considered during the last decades.

Nearly all countries have faced major floods in the last decade. The Netherlands faced critical situations in 1993 and 1995 and flooding of some polders in 1998.

France, Italy and Austria share the problem of steep rivers and flash floods. Nearly all countries share the problem of flooding of lowland rivers – with or

Table 5.1. Questionnaire of quick scan.

1. Types of flood problem
   - Types of flood problem (coastal flooding; large rivers; small rivers; local rainfall; urban floods)
   - Extent of damage

2. Political attention for flood problems
   - Political momentum
     - Is flooding an important policy issue?
     - Are new policies and legislation being prepared?
   - Organisation
     - Who is responsible for flood management?
   - Sense of direction
     - Is a relation made with climatic change?
     - Is the orientation towards space for water or towards more technical solutions?)

3. Solutions: what kind of measures, apart from technical measures such as dykes, pumps, reservoirs?
   - Retention in the catchment (forests, flood meadows, meandering brooks)
   - Retention in urban areas (decreasing impervious areas, decoupling roofs)
   - Retention along rivers (storage areas for flood water)
   - Floodplain restoration (removing obstacles, (re)introducing river channels, dyke setback)
   - Are these measures combined with other objectives, such as environmental restoration, recreation, new housing areas, landscape reconstruction

4. Solutions: what kind of instruments?
   - Hazard zoning and building regulations
     - Is hazard zoning part of national, regional and local plans?
     - Is it obligatory or not?
     - What categories of zones are used?
   - Role of insurance and compensation schemes
     - Is flood damage compensated by the state?
     - Is commercial insurance against flood damage available?
   - Role of economic and fiscal instruments
     - What incentives are available for preventing flood damage?
     - What incentives are available to increase retention?
   - Role of financial arrangements
     - Are decisions made on the basis of a cost-benefit analysis?
     - Are there possibilities for co-financing with a view to environmental restoration, recreation?
     - What is the role of external subsidies, such as EU funds?
     - Is there co-financing by private funds?
   - Role of planning procedures
     - What is the status of plans and planning procedures?
     - Who is involved?

5. Useful institutes, projects and key persons for further contact

Steep rivers   flashfloods and mudflows

Lowland rivers without dykes   floods over restricted areas

Lowland rivers with dykes   floods over substantial areas

Elevated rivers with dykes   pontentially huge floods and large drainage problems

Polders, flat and poorly drained   floods due to excess local rainfall

Estuaries   floods due to sudden storm surges

Table 5.2. Main flood types in selected EU countries.

	NL	B	D	UK	F	HU	PL	I	A	S
Recent large floods	1993–1999 (*)	1995	2002	2000	1995	2002	1997	1996	2002	
Steep rivers		X	X		X		X	X	X	
Lowland rivers without dykes	X	X	X	X	X	X	X	X	X	X
Lowland rivers with dykes	X	X	X	X	X	X	X	X	X	
Elevated rivers with dykes	X					X				
Polders, flat and poorly drained areas	X	X	X	X	X	X				
Estuaries	X	X	X	X	X					

Netherlands (NL), Belgium (B), Germany (D), United Kingdom excluding Scotland (UK), France (F), Hungary (HU), Poland (PL), Italy (I), Austria (A), Sweden (S).
(*) near-floods of major rivers occurred in 1993 and 1995; in 1998 and 1999 various polder areas were flooded due to excess rainfall.

without dykes. The problem of flooding of elevated rivers is mainly faced by The Netherlands and to a lesser extent by Hungary.

5.3.2   Political momentum

Recent floods or near-floods have put flood risk prevention high on the political agenda in all countries that we included in the quick scan; with the exception of

Table 5.3. Standards for flood hazards in selected EU countries (unit: 1/year).

	1:50	1:100	1:200	1:500	1:1000	1:2000	1:5000	1:10000
Steep rivers								
Lowland rivers without dykes		HU, FR, D, I	D					
Lowland rivers with dykes		HU, FR, D, I, UK	D		NL			
Elevated rivers with dykes					NL			
Polders, flat and poorly drained areas	NL	NL						
Estuaries			UK	B			NL	NL (B)

Notes: Values have not been validated. The figures are flood hazard standards, not actual flood hazards. Protection levels along urban areas in HU, FR, D, I, UK are often higher – typically 1:500. In the UK there are no legally established standards. In Belgium the actual flood hazard along the Scheldt is 1:350 years; a project is in progress that will decrease this level to 1:10000 years.

Sweden. In all these countries policies are being re-evaluated in order to face this risk. There is strong interest in the application of spatial measures to reduce flood hazard and in the reduction of flood damage potential. Poland may be the exception, as the focus still appears to be on technical measures for the reduction of flood hazard.

An issue that is attracting increasing attention, is the continuous increase in the flood damage potential due to investments in areas with a non-negligible flood hazard, for example in areas behind dykes with a 1:100 year standard. (For the perspective: a 1:100 year flood has a 26% probability of occurring once in 30 years, the duration of a typical mortgage).

In most countries the power for spatial decisions is delegated to local authorities. In many countries such as England, France, Germany and Austria central authorities are taking initiatives to sensitise or influence local authorities to take flood hazard and flood damage potential into account in their spatial decisions.

In England, Germany and The Netherlands climate change is seen as an important factor contributing to an increase in flood risk.

Along international rivers with flood problems, the level of international co-operation is increasing. This is the case along the Rhine, the Elbe and the Oder. The Interreg IIIB programme is a focal point for international co-operation, as it includes many projects in the field of flood risk prevention in the region of North West Europe and to a lesser extent in the North Sea Region, Central and South-Eastern Europe (CADSES) and the Alps.

Table 5.4. Main measures applied in selected EU countries.

	NL	B	D	UK	F	HU	PL	I	A	S
Reducing flood hazard										
Technical measures										
Dams, dykes, pumps, reservoirs	R	R	R	R	R	R	R	R	R	r
Spatial measures										
Retention in the catchment										
Restoration of forests, flood meadows and meandering brooks	Pr	p	R	Pr	Pr			p	R	r
Retention in polders and flat areas										
Increasing area of surface water, flood meadows	Pr									
Infiltration in urban areas										
Decreasing impermeable areas, reducing the area drained by sewers, de-coupling roofs	Pr		R	Pr	P			p	P	r
Retention areas along rivers										
Measures aimed at storing flood water	Pr	R	R	Pr	Pr	R	pr		R	r
Floodplain restoration										
Measures aimed at increasing the river flow capacity	Pr	Pr	Pr	Pr	Pr	Pr	p		Pr	r
Reducing flood damage potential										
Flood proofing houses and infrastructure	r		R	R	r				R	
Removing houses		r			p	r				

5.3.3   Measures

The information on the measures in the 10 countries has been scored on a 4-point scale:

p – tentative plans
P – definite plans
r – realisation on a small scale
R – realisation on a large scale

Results are presented in Table 5.4. The scoring procedure is subjective, and depends on the completeness of the information per country, and may therefore not necessarily reflex the general situation/policy per country. The table and the discussion based on it may contain errors, as can be expected from a quick scan.

5.3.3.1   *Technical measures*
Technical measures, which are not the focus of this report, still form the backbone of flood prevention in all selected countries.

5.3.3.2   *Small reservoirs*

The creation of systems of smaller reservoirs to attenuate floods is popular, especially if this can be combined with water provision in summer. Aquatic recreation is also a motive. Most examples are found in southern France and Italy, in regions that face long dry summers. But there is also renewed interest in using existing, often hydroelectric reservoirs for flood management purposes, for example in Germany and France. There are often conflicting interests, as constant high water levels are preferable in recreational areas. There is a long tradition in multipurpose reservoir management in the US, but not so in Western Europe.

5.3.3.3   *Spatial measures*

In all countries considered (except Sweden), recent flood events have placed spatial measures on the political agenda. Table 5.3 gives some insight into who are the "leaders" and "followers", but in general most countries appear to be at approximately the same stage of development, that is between definite plans and realisation on a small scale.

5.3.3.4   *Retention in the catchment*

There is much attention for retention in the catchment in Germany, Italy, Austria and France.

– In the catchment areas of the larger German rivers – the Rhine, Oder and Elbe – this has been an ongoing activity for many years. In the last 5 years particularly, a large number of projects have been implemented.
– Austria is attempting to integrate the planning requirements of the WFD and the planning of catchment restoration in order to prevent flash floods and erosion.
– In Italy river basin authorities address the issues of soil erosion, slope stability and flow capacity in their river basin plans; these issues lie within the competence of the river basin authorities.

In steep catchments, measures focus on reforestation, slope stability and enhancing stream flow capacity. Environmental restoration is often an important additional goal.

In hilly catchments, retention focusses on reduction of agricultural drainage, re-meandering of brooks and restoration of riparian lands. Sometimes the restoration of groundwater levels and upward seepage areas are additional objectives. In Germany there are specific guidelines that demand the restoration of the natural hydrology in order to achieve ecological objectives.

Most of these approaches are limited in scale and focus upon smaller catchments, but are not seen as an alternative strategy for managing large-scale flood problems. Many of the measures are subsidised by national nature conservation budgets and by EU money. With the WFD the reduction of diffuse pollution by agriculture will be high on the agenda in all catchment plans. The combination of the needs of water retention and emission reduction offers a challenge.

5.3.3.5   *Retention in polders and flat areas*

In polders in The Netherlands, initiatives are being taken to enlarge the retention capacity of drainage systems (by widening channels and ditches and by designating the lowest grasslands as inundation areas), and applying dynamic forms of water level management.

Outside The Netherlands, no examples have been found.

5.3.3.6   *Infiltration in urban areas*

Plans for infiltration in urban areas abound, but implemented plans are still scarce.

- In Germany, retention in urban areas receives much attention. Infiltration is mostly combined with new housing areas. Rain water infiltration is often implemented when sewers have to be renewed. A hydrological guideline prescribes that the 1-year maximum runoff (HQ1) should not differ more than 10% from the original HQ1 under near natural conditions. The HQ1 reference is calculated on the basis of a catchment area in which all urban areas have been replaced by a representative combination of forests and agricultural land.

- Austria is in the process of converting combined sewage system to a separate sewage system, with infiltration of the non-polluted runoff via soil filters into permeable soils.

- In England measures such as infiltration, retention and buffers are increasingly being considered in urban areas through the use of SUDS (sustainable urban drainage systems). The use of buffer zones within urban areas is a favoured option but it takes time to realise the release of currently developed areas for this purpose.

- In Italy the river basin plans often contain the requirement of hydrologic compensation in case larger surface areas become impermeable due to urbanisation.

5.3.3.7   *Retention areas along rivers and floodplain restoration*

Retention areas can be found in most countries. Many examples exist in Germany, Italy, Belgium and Hungary.

- In Germany along the Rhine and the Elbe many retention areas are in operation. An even larger number of retention areas is being planned and constructed. The size of these areas is typically 100–600 ha. The Havel Polder is a 2000–4000 ha retention area that is in operation along the Elbe.

- Hungary has large retention areas (2000–5000 ha) along the Tisza, that have been operational during the recent floods.

- In Italy retention areas (200–500 ha) have been constructed since the beginning of the 20th century; many new initiatives are in the planning stage.

- In Belgium the Sigmaplan is an ambitious plan to reduce flood risks along the Scheldt, by means of the creation of a large number of – relatively small – retention areas.

- There is growing interest in *floodplain restoration* along larger rivers. Examples can be found in Austria, Belgium, Germany, France, Hungary, Italy and England.

- In Germany dyke setbacks are an important element of the recently adopted Flood Action Plan for the Elbe.

The above countries have extensive plans for new retention areas and floodplain restoration. Generally, the areas are situated in sparsely populated regions and are combined with the objective of environmental conservation, which can give the legal title and funds necessary for land acquisition. Other common functions are forestry and extensive forms of recreation. Combination with agriculture and intensive forms of recreation are less common. We found no examples of combination with housing.

Creating new retention areas and regaining floodplains is much more difficult than maintaining existing areas. Many new retention areas are still on the drawing board. Creating the necessary local coalition for actual realisation of these plans is generally difficult. There are frequent instances of local resistance, for example in Germany, Belgium, Italy and Hungary. Success stories are characterised by a win-win approach that cater for the wishes and requirements of local stakeholders and involve tailor-made designs that fulfil multiple objectives and open the door to multiple budget lines. Spatial solutions may have the most societal effects with the least costs to the local authorities and inhabitants. The reverse side of the medal is, that most success stories appear to be based on very generous external funding at a national level or from the EU.

5.3.3.8 *Reducing flood damage potential*
Reducing flood damage potential is central to the flood management strategy in Austria, Germany, France, Italy and England.

Reducing flood damage potential mainly employs instruments (**see**). The most obvious (physical) measure for reducing flood damage potential is flood proofing, which is often prescribed via building regulations (**see**). Flood proofing is being developed and applied on a considerable scale in Germany, France, England and Italy.

Removal of houses, the ultimate retreat from flood-prone areas, requires juridical and economic instruments such as expropriation and state compensation (**see**).

5.3.4   Instruments

An overview of the instruments that we found in the quick scan. We used the same scoring procedure as for the measures. It must be stressed that the table has not been validated, so an empty cell must not be read to mean that the instrument is not used in the country concerned.

We assigned the instruments to one of three categories: juridical, economic and communicative. This is a common distinction but only partly satisfactory. All instruments have indirect effects that could warrant their assignment to one of the

Table 5.5. Instruments applied in EU countries to achieve space for water.

	Effect	NL	B	D	UK	F	HU	PL	I	A	S
Juridical instruments											
Obligation to consider flood risks in land use decisions	DP	R	R		R						
Restrictive hazard zoning	DP	r		p	Pr		R	pr	pr	R	R
Building regulations (e.g., flood proofing)	DP			Pr			R	pr		R	
Reservation of floodplains and retention areas in spatial plans	H		Pr	P	R		R			R	
Expropriation because of flood risks	H, DP			r			r				
International treaty	H				R		R				
Economic instruments											
Subsidies for the implementation of spatial measures	H		r	R	R	R					
Private contracts to accept floods	H		r		r		r	P			
Tax on impermeable surface in urban areas	H			R			P				
Private insurance of flood damage	DP		p	P	R	R	p	p	r	R	r
State compensation for flood damage	?		r	P	r		R	R	r	R	
Communicative instruments											
Flood hazard mapping	DP			p	R	R	R	p	pr	R	r
Complementary instruments											
Land reallotment schemes used to implement spatial measures	H		p		R		r				

NB.: empty cell does *not* mean that instrument is not used in the country concerned.
H: instrument in support of measures to decrease flood hazard.
DP: main effect of instrument is the reduction of flood damage potential.

other categories. For example a prohibition to build in a flood-prone area is a juridical instrument, but also has economic effects (land prices will drop due to the prohibition) and communicative aspects (people realise that an area is danger-ous). In addition, instruments are chosen to implement or facilitate specific measures, thus per measure a classification scheme of instruments could be devel-oped. We tackled this aspect in part, by giving each instrument one of two labels: instruments in support of measures to decrease flood hazard (H) and instruments that have their main effect by decreasing flood damage potential (DP).

5.3.5   Juridical instruments

5.3.5.1   *Obligation to consider flood risks in land use decisions ("Water Assessment")*

In the Netherlands, Belgium and England the national authority has obligated the local authorities, who take the land use decisions, to consider flood risks explicitly in their decisions, or to ask the advice of central authorities with expertise on flood risk reduction. However, the power of decision taking has been left with the local authorities, and the advice of central authorities does not need to be followed. Thus, this can be regarded as a process instrument, as it impels that a process should be followed, but does not prohibit decisions that increase flood risks.

5.3.5.2   *Restrictive hazard zoning and building regulations*
In many countries restrictive hazard zoning, in conjunction with building regulations, form the core instruments of flood risk reduction.
- In France, Italy and Austria the designation of hazard zones and the regulation of building activities in these zones is required by law.
- In Germany the recent Flood Prevention Law (*Hochwasserschutzgesetz*) obliges the Lander to apply restrictive hazard zoning
- In Germany, France and Italy there is increasing attention for hazard zoning of areas with a high capital investment that are protected by dykes.

Restrictive hazard zoning and building regulations are instruments that are generally used in conjunction. The hazard zones delimit the zones with a certain flood hazard (often 1:100 years); the building regulations specify the necessary measures for flood damage reduction in these zones. Common elements are:
- to guarantee human safety (prevent the collapse of buildings and bridges, keep escape routes operational, prevent buildings with only a ground floor in areas that are deeply flooded)
- to safeguard critical public services (power stations and hospitals to be planned outside flood prone areas).
- to limit flood damage (flood proofing of ground floors, forbid the construction of cellars)
- to limit environmental damage (prohibition of oil tanks, flood proofing of potentially polluting activities such as chemical industries).

In contrast with the "Water Assessment" described above, these instruments attempt to prohibit decisions that increase flood risks.

5.3.5.3   *Reservation of floodplains and retention areas in spatial plans*
In most countries, maintaining "empty space" for water functions is a challenge, as the pressure from other functions is strong. Thus, it is necessary to give floodplains and retention areas an explicit protected status in spatial plans.
- Germany has a long tradition of protection of floodplains and retention areas. However, the protection status can differ per province. With the recent Flood Prevention Law, the federal government has attempted to bring the protection status of these areas onto a common footing, and to stimulate the designation of new areas.

- In France, retention areas and floodplains can easily be protected in the framework of the restrictive hazard zoning plans (*Plans de Prevention de Risques*).

– In Italy the retention areas have a clear protected status.

– In The Netherlands the floodplains between the dykes are reserved for water discharge. Also inland lakes and waterways are delimited and protected from land reclamation.

Reserving space for water for the future is an even larger challenge. This requires a change in land use status in the land use plans.
– At present, Belgium is taking this step as part of the Sigma Plan.

5.3.5.4   *Expropriation because of flood risks*
Only a few instances of the application of expropriation have been found.
– In Belgium the implementation of the first phase of the Sigma Plan led to the expropriation of some isolated houses in future retention areas. The instrument may become more important as the retention areas in the second phase of the Sigma Plan have a more intensive land use. Expropriation can work both ways: inhabitants that live in flood prone areas can also offer their premises to the state, who is obligated to buy them.

– In France the Barnier Fund finances voluntary expropriation of property when there is an acute flood hazard and the costs of preventive measures for public authorities exceeds the value of the property.

5.3.5.5   *International treaty*
In 1982, the national governments of France and Germany agreed on a program of restoration of floodplains and realisation of retention areas along the Middle Rhine. As the agreement took the form of an international treaty, it provided powers to the national authorities to impose regulations on the local authorities and owners.

5.3.6   Economic instruments

5.3.6.1   *Subsidies for the implementation of spatial measures*
Considerable subsidies from higher levels (province, national, EU) are widely used and appear to be a pre-requisite for the implementation of spatial measures.

5.3.6.2   *Private contracts to accept floods*
The land that required for floodplains or retention areas is almost always privately owned. This land may remain in the hands of private individuals if contracts guarantee their proper "blue" use. Examples of compensation payment for a "blue" use of land were found in France, Germany and The Netherlands. Preferably, the regulations that safeguard the water function are not in the form of a contract with a temporary land owner, but are vested in the land, for example in the form of an easement.

5.3.6.3   *Tax on impermeable surfaces in urban areas*

This type of instrument has been found in Germany and is under discussion in France.

- In Germany, many communities (e.g., Berlin) have introduced a stormwater fee for discharging stormwater into the sewage system, while simultaneously reducing the sewage fee based on drinking water consumption. The introduction of this system, which is called a "split fee" is a rather complicated task. The sealed area must be surveyed by areal photos, the degree of connection must be estimated and a database with all the information including the addresses of the owners must be constructed. The process is usually accompanied by a lot of information (web, letters, public hearings, etc.) and also legal procedures (changing the local by-laws). Due to the "split fee", the awareness of private and commercial property owners for stormwater issues is increasing.

- In France, a similar system is now under discussion. It is proposed that *Agences d'Eau* levy a tax for all impervious areas, in order to create a fund that can be used for floodplain restoration.

5.3.6.4   *Private insurance or state compensation of flood damage*

In nearly all countries there is a system for the compensation of individuals who suffer damage from large floods.

- France is an example of a country with state compensation and a poorly developed system of private insurance. Flood damage is compensated by a Natural Catastrophes Fund that is financed by an automatic deduction on all mandatory household and vehicle insurance policies. This deduction has increased from 9 to 12% in recent years, and now amounts to approximately 1 billion Euro per year. Local Mayors can apply for compensation when a catastrophic event occurs and increasingly do so. The insurance companies manage the fund. This arrangement is under scrutiny of the European Commission, because it breaches the principle of free choice of insurance[6].

- In contrast, in England a strong insurance industry provides (voluntary) insurance on a commercial basis. There is no state damage compensation scheme and government explicitly states that it will not compensate flood damage. This is believed to be a pre-requisite for the successful functioning of the private insurance scheme, as individuals will not choose private insurance if they can count on state compensation.

- Mixed situations also occur, where private insurance pays those who are insured, and the state is obliged to – partly – compensate uninsured parties who have suffered losses during large floods.

---

[6] For the same reason, a scheme for compulsory flood insurance in Baden Wurttemberg – Germany had to be cancelled.

– In Belgium the discussion between private insurance or state compensation of
  flood damage is presently on the political agenda.

Because of the high costs incurred in recent floods that had to be covered by
national authorities and the EU, there is increasing interest in private insurance.
Private insurance is often combined with restrictive hazard zoning and building
regulations.

5.3.7   Communicative instruments

Few cases were found of instruments with a primarily communicative function.
This is partly due to the fact that this question was not explicitly asked in the questionnaire. However, from an extensive search of the internet, it appears that this
type of instrument is not very popular.

5.3.7.1   *Flood hazard mapping*
Flood hazard mapping is the basis of restrictive hazard zoning and, as such, countries that use restrictive hazard zoning also have flood hazard maps. In England,
where there is no obligation for local authorities to apply restrictive hazard zoning,
detailed maps of flood hazard are freely available on the internet, with the aim of
informing the local authorities and the public of the flood hazards that they face.

5.3.7.2   *Complementary instruments*
In France and Germany land reallotment is an important complementary instrument especially if one wants to create a win-win solution with agriculture, which
is often the most critical stakeholder and landowner in flood plains. A land reallotment scheme puts land ownership on a fluid basis and can increase external
finances that facilitates the process.

5.3.8   Specific points of interest per country

In *Austria* a consensus is growing that local spatial planning by municipalities is
no longer sufficient to tackle flood risks. There are initiatives to make flood risk
reduction plans for whole river basins. As the requirement for river basin planning
is also set by the EU Water Framework Directive, there are efforts to integrate both
types of planning.

   In addition, a discussion is going on concerning the financing of spatial
measures for flood hazard reduction. Downstream municipalities are benefited by
measures taken by upstream municipalities; how can the costs be shared between
the two?

In *Flanders* there are interesting features in the spatial planning system.
– There is a governmental right of pre-emption in areas designated as such in its
  spatial plans, that may be used to realise retention areas.

– There is compensation for planning damage if, due to a change in the spatial plan, a location loses its right to realise housing; 80% of the damage is compensated by the government.

In *England* the market and economic approach to flood risk reduction is most advanced.

– In contrast with other countries, precise data on the flood damage potential are readily available.

– Real estate developers must provide an assessment of whether any proposed development is likely to be affected by flooding and whether it will increase flood hazard elsewhere.

– The developers must fund the provision and maintenance of flood defences that are required because of the development.

– Grants from the national government budget for flood prevention measures are provided on the basis of an assessment with a detailed benefit/cost analysis. In the benefit/cost analysis the benefit of prevented flood damage and the cost of flood prevention measures are compared at various levels of flood frequency, in order to find the optimum set of measures. It is required that the managed retreat as well as abandonment of existing defences be considered in each project. In addition, social factors and environmental acceptability are important in the assessment.

While in Germany there is much attention for spatial measures, in Poland the emphasis is still on technical measures. Thus, strategies implemented along the east bank of the oder may differ from those used along the west bank.

In Sweden flood risks are so small that the emphasis is on efficient emergency and rescue procedures if they do occur.

6

Risk Perceptance and Preparedness and Flood Insurance

Paul Baan
WL/Delft Hydraulics, Delft, The Netherlands

ABSTRACT: Technical and economic analyses are often made to determine the feasibility of measures/projects as part of a flood protection plan. Mostly, institutional and administrative aspects are also addressed. But how the people living in the polders feel about the plans often gets next to no attention. This paper is devoted to the attitude of the citizens focusing on the situation in the Netherlands. How are people living with (flood) risks and how do they feel about it? What are the needs for communication on risks, and is compensation of flood damage recommendable?

6.1   FLOOD PROTECTION IN THE NETHERLANDS

Protecting polders which lie alongside rivers from flooding has been effective in the Netherlands: the last flooding of polders in the Netherlands occurred in 1926. In that year, the discharge level of the river Rhine was the highest ever observed. Extreme flood waves also occurred in 1993 and 1995. In 1995 the polders along the Rhine river had a narrow escape from flooding. The extreme Rhine river discharges in 1993 and 1995 raised flood risk awareness once more among the people and government in the Netherlands. Since then levees have been heightened and strengthened.

   In the 1990's people and government started worrying about the technically based flood protection philosophy in the Netherlands. By heightening and strengthening the levees the Dutch can and did counteract the increasing frequency of extreme river flood waves due to climate change and increased run-off in the river basin as a result of increasing urbanisation. But with the increasing likelihood of extreme water levels in the rivers and growth of the population and the economy in the low-lying polders, the impacts of an eventual flood might become disastrous. Also taking into account that absolute protection against flooding cannot be offered, in the mid-nineties the "room for the river" concept was adopted by the government. The protection level is maintained by giving more room for the rivers aimed at reducing the extreme water levels. In this way the Dutch can keep up with increasing extreme flood waves. Moving levees inland

and removing obstacles in the river bed to prevent damming up of water are possible measures. The government in the Netherlands is also considering the assignment of calamity polders that will be inundated on purpose at times of extreme river water levels to protect other polders from flooding. Levees are no longer heightened, except locally, when there are no other solutions.

The extreme water levels in the river Rhine in 1993 and 1995 and the elaboration of the "room for the river" concept give rise to disconcertion among the inhabitants affected, i.e., the people living in the potential calamity polders feel very uneasy about the plans and protest.

In decision making on flood protection, technical and economic analyses are made to determine the feasibility of measures comprising part of a flood protection plan. Usually, institutional and administrative aspects are also addressed. But how the people that are living in the polders feel about the plans often gets less attention. How the people feel about living along the rivers and their concerns about flood risk and their personal consequences of a flood should also be an important issue in decision making on flood protection.

6.2 LIVING WITH RISKS

6.2.1 Introduction

Risk is generally defined as the chance of an unwanted event times the consequences (damage and losses due to that event). The choice of acceptable risk levels and the degree of variation in it is a political issue. Risks can be compared based on chance times effect. Subsequently, a cost effectiveness analysis (how much risk reduction can be realised at what price) is possible to determine whether money is spent well (RIVM, 2003). However, such risk trade-offs do not confront the basic opportunity cost problem. The key question is of where our priorities lie; is it better in terms of social welfare to invest in reducing flood risk, or in improving public health, water security, or food security or in reducing ecological risks? (Rees, 2002).

In judging risks, citizens take more aspects into account than the results of a cost effectiveness analyses. Qualitative aspects, such as the perceived degree of free choice, equity, degree of control of the risks and gains that can be expected, are often decisive for risk acceptance (RIVM, 2003). People also base their judgement on feelings. When people like an activity, they underestimate the risk and overestimate the possible gains, and the reverse (Slovic et al., 2002).

Human actions, processes in nature and combinations of both all involve risks. Risk is an integral part of human life. We cannot and do not live in a risk free society; the taking of risks has been the engine of economic and social development (Rees, 2002). Due to technological development, the risk of large and small disasters has decreased in the last century. Evidence of this is improved health and a longer life expectancy. However, the remaining risks receive much attention.

This holds also for floods. During the last centuries, the risk of flooding has been decreased considerably in the Netherlands by constructing, heightening and strengthening a solid network of levees. In policy making, further decreasing the flood risks is an important issue. Recently, many studies have been carried out and discussions are going on concerning the room for the river concept and the assignment and use of calamity polders (see, for example, Dijkman et al., 2003).

Increasingly, citizens in the Netherlands want to decide themselves on the risks they take. They want to be free to take self-chosen risks, but they also demand absolute protection against risks they do not want to take. From a calculating perspective, the citizens in the Netherlands expect that government income generated from taxes be used to protect them against risks which are not self-chosen. Or stated otherwise, citizens have the idea that they are insured by government against unwanted risks (RVZ, 2001).

6.2.2   Dance of effect and reason

There are clear differences in the way the public and the government (and risk experts) view risks. In evaluating risks, government and experts apply an analytical approach emphasising the use of quantitative data while the judgement of the public, when involved, is based on experiences and feelings. The analytical approach uses algorithms and normative rules, such as calculation of probabilities and formal logic, to determine risks. The public, when involved, reacts more emotionally based on earlier experiences and makes judgements mainly based on qualitative aspects and feelings (Slovic et al., 2002; Flinterman et al., 2003; Slovic & Weber, 2002). However, water risk experts also cannot be depicted as neutral, disinterested protectors of the public good. Trying to maintain jobs, budgets or research grants, and the aggrandisement of bureaucracies will all influence their input to decision making (Rees, 2002).

Slovic et al. (2002) state that an analytical approach to risks can only be effective, when it is guided by emotion and feelings (the dance of effect and reason). People base their judgement not only on what they think about a topic, but also on what they feel. Thus, feelings have a direct effect on the judgement of risks. According to Slovic et al. (2002), scientific research on rationality and feelings is still in its infancy.

Figure 6.1 presents an overview of the way of viewing risks by government and experts (the technical rationale) on the one hand and the public (cultural rationale) on the other.

6.2.3   Risk perception and acceptance

The perception and acceptance of risks is dependent on the socio-cultural context, the characteristics of a risk (e.g., man-made or a natural cause), the degree of free choice, the extent of control we have over the risk and its consequences and the

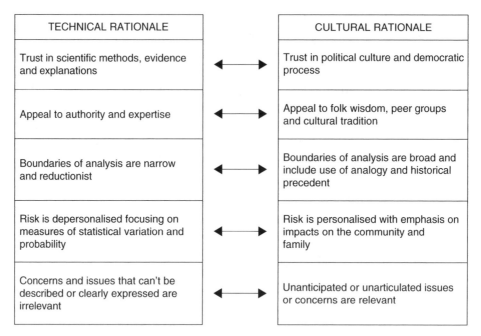

Figure 6.1. Technical versus cultural rationality in viewing risks.
        Source: Barnes (2002)

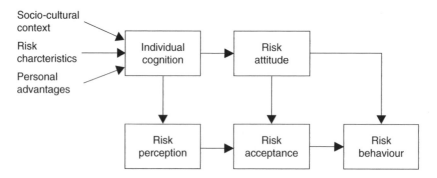

Figure 6.2. Scheme of individual cognition of risks and resulting risk behaviour.
        Source: adapted from Flinterman et al. (2003)

question of how much personal advantages is at stake. Figure 6.2 presents a schematic overview of the individual cognition of risks and the resulting risk behaviour.

Many researchers have been involved in determining the factors (dimensions) of risk that can explain differences in risk perception and acceptance. The number of factors found varies from a selection of a few important ones or some clusters to a long list. Sjöberg and Drottz-Sjöberg (1994) drafted a more or less complete list of 27 factors, of which 10 are related to the character of the risk, 11 to the social

Table 6.1. Scores of five different risks on the risk factors according to Lyklei.
of 0 (unimportant) to 4 (very important or serious).

Risk factor	Smoking	Beesting	Nuclear power plant	Road traffic	Flo.
Chance of death	10^{-3}	2×10^{-7}	10^{-7}	10^{-4}	10^{-7}
Fairness of division of risk and gains between parties	0	1	4	1	1 (4)
Familiarity with risk and effects	0	1	3	1	2
Reasons of exposure to risk	0	1	3	0	2
Seriousness of effects	1	1	4	2	3
Degree of preparedness and of control of consequences	2	1	3	1	2
Percentage of maximum score	15%	25%	85%	25%	50%
					(65%)

* When scores for inundation differ from those assigned to (uncontrolled) flooding the scores are presented between brackets.

context, and 6 to personality features. Personality features are important to determine individual risk perceptions. For instance females judge risks mostly more serious than males (Weber et al., 2002). The insight in individual risk perception is far from complete. Currently, all factors together explain less than half of the differences in individual risk perception (Sjöberg, 2000).

Lyklema (2001) carried out research to find the key factors for risk in water management in the Netherlands. It turned out that the following factors are most important:

- the fairness of the division of risks and gains between parties involved;
- the familiarity with the risk and the effects on the people exposed;
- the reasons of the exposure to the risk;
- the seriousness of the effects;
- the degree of preparedness of citizens and the degree of control over the consequences.

Vlek (2001) emphasises the importance of Lyklema's fifth factor stating that the perceived control is a key factor for understanding the judgement, choice and behaviour of people in risk situations.

In Table 6.1, for a few quite different risks, the author assigned scores to the risk factors. Though this way of scoring is subjective and the uncertainty margins are relatively large, rather robust conclusions may be drawn from the results. The risk of a disaster with a nuclear power plant seems most threatening. Scores for flooding are also rather high. Both represent involuntary risks with hardly any individual gains. Inundation is more threatening than (uncontrolled) flooding. Risks of smoking and road traffic are clearly viewed as less of a risk than that of a

ar power plant or a flood. Smoking and participating in road traffic are vol-
ary activities with individual gains. Moreover, the impacts are mostly small.
he risk of a bee sting is considered a natural and small risk.

Media and social processes influence the perception and acceptance of risks.
When the media are the only source of information for the public, the effect on the
risk perception may be great. Media can contribute to what Slovic and Weber
(2002) call "social amplification of risk". An important aspect of social amplification is that the direct impacts need not be too large to trigger major indirect
impacts. A small incident in an unfamiliar system (or one perceived as poorly
understood) may have immense social consequences if it is perceived as a harbin-
ger of future and possible catastrophic mishaps. Media jump on it (consider BSE,
SARS, bird flu), as such events have a high "signal value". As a result efforts and
expenses beyond that indicated by a cost-benefit analysis might be warranted to
reduce the possibility of "high-signal events" (Slovic & Weber, 2002; RIVM, 2003).

Strong social networks influence individual perceptions. Individuals who are
most connected to each other through interpersonal contact are also most likely to
share similar information, attitudes, beliefs and behaviours on controversial topics.
Analysis performed by Scherer and Cho (2003) confirmed that social linkages in
communities may play an important role in focusing risk perceptions. A collective
attitude based on a social network may result in the formation of a "socio-
psychological shield", which prevents information from coming through. With
strong social networks attempts of government and experts to change attitudes,
beliefs and behaviour may have little or no effect.

6.2.4   Communication on risks

The complex nature of risks and the often great impacts make huge demands on
communication. People deal selectively with information; they take in only what is
relevant and convenient to them and what fits with their idea of reality. Former experiences and social networks are important too. This means that people, when threatened with a flood, do not always react as the government would expect or like to see.
For instance, it is difficult to get everybody evacuated in time from an area threatened by flooding (Slootweg & Schooten, 2002).

One-way communication on risks from government to the public fails. Instead,
the people and parties concerned should be involved at an early stage in the policy
making and decision making process for risks and the communication of it. This
enables people to contribute to thinking about the problems, the analysis of possible
solutions and the conclusions. Involving the people in the analytical and decision
making process contributes to gaining a sound social basis. Arvai (2003) proved
this with experiments on participative decision making.

People wish to be informed in a language intelligible to them and they also want
their feelings to be taken into account. Moreover, in communicating, government

must be clear on uncertainties, as Stefanovic (2003) states: sounding more certain than the data can justify is a sure way to lose trust and credibility.

People understand percentages of probability better than frequencies (Slovic et al., 2002). Expressing risk impacts in a positive way (number of survivors) seems less threatening to people than expressing these impacts in a negative way (number of deaths). Currently, flood risks are often expressed as frequencies, e.g., once in every 500 years. Expressing flood risks in a positive way and applying percentages of probability is possible. For example, a flood frequency of once in 500 years can be presented as a protection level of 99.8% a year. Psychologically, this way of expressing risks has more advantages. It clearly states that: (a) the protection level is below 100% meaning that absolute protection cannot be warranted; and (b) often expensive flood protection measures may improve the protection level by only very little.

6.3   LIVING WITH THE RISK OF FLOODING

6.3.1   Worrying about floods

Among all kinds of risks, risks of flooding clearly are not self-chosen by men. Inhabitants of low-lying polders along the rivers in the Netherlands, therefore, expect absolute protection. A solid levee network must warrant this protection. As a result, people living in these polders in the Netherlands worry little about the risk of flooding. Environmental psychologists explain this by what they call "the crisis effect", and "the levee effect".

The crisis effect indicates that attention to a disaster is greatest during and immediately following its occurrence, but awareness dissipates greatly between disasters (Stefanovic, 2003). People are short of memory. Shortly after a (near) flood people overestimate flood frequencies. After some time, the worries decrease. Some years later flood frequencies are again underestimated (Penning-Rowsell, 2003).

The levee effect refers to the fact that once measures are taken to protect against a disaster, people place unrestrained and often inappropriate faith in the power of the technology to protect them. People seem to be lulled into thinking the levee will protect them from all future floods (Stefanovic, 2003) and live light-heartedly behind the levees. As a consequence, people living in the polders do not feel the need to anticipate a possible flood and are ill-prepared for floods. Also, as people living in the polders have no experience with flooding, they fear the consequences of an eventual flood.

6.3.2   Threat of flooding

Due to the large impacts of the flooding of low-lying polders along the rivers in the Netherlands, flood wave events are threatening to the people living in these polders. People living in the river meadows are used to the dynamics of river water and feel less frightened than people living in the polders. People often find evacuation more severe and threatening than the high water levels in the river (Vlieger et al., 1998). New dwellers in the polders are less familiar with water dynamics and its related flood risk and often are more afraid and feel more stressed than original dwellers. The original dwellers have developed knowledge and skills to cope with situations of high water levels and are better prepared. Experiences with high river water levels and evacuation make flood risks imaginable and less threatening (Slootweg & Schooten, 2002).

Some groups of people are dependent on care from others and as a result are more vulnerable to flooding: the elderly, sick and invalid people and those living in isolated places. These vulnerable groups find risks more threatening. When evacuation is needed these people need special attention (Slootweg & Schooten, 2002).

Feelings of inconvenience and stress start with rising water levels in the river and when there is a chances that extreme levels may be reached. When evacuation is needed the feelings of stress are aggravated. A long prolongation of the evacuation further increases the feelings of stress. The attitude of people to the stress and feelings of inconvenience caused by high water levels in the river differ strongly between individuals. Some dwellers remain easy-going while others feel very threatened (Baan & Rabou, 2002).

Positive feelings occur also. People in a polder threatened by high water levels in the river have positive feelings related to togetherness and solidarity. Social coherence and the relationships between inhabitants improve in such periods. Some people even feel the threat of high water levels as exciting and like it (Baan & Rabou, 2002). Gratitude for mutual help is also mentioned as a positive effect (Slootweg & Schooten, 2002).

6.3.3   Flood preparedness

Exchange of information about experiencing natural disasters increases risk awareness and helps people in preparing for a possible disaster. However, gaining experience is a lengthy process and cultural adaptation is also involved (Gebre & Bruen, 2002). Citizens, who are well prepared and can react adequately during and after a flood, are less affected psychologically and feel less threatened. Communities of well-prepared people find flooding better controllable and such communities are socio-economically less vulnerable (Baan & Klijn, 2003). In Germany it was proven that experience with flooding and as a result improved preparedness resulted in improved control over the flood: the economic losses from a flood event in 1995 were only half of that from an earlier one in 1993, although the two events were of comparable size (Kron & Thumerer, 2002).

The lower the frequency of flooding, the stronger the tendency of the threatened people and businesses to shift the responsibility for protection and preparedness to public authorities (Kron & Thumerer, 2002). When frequency is low, it is difficult to increase flood risk awareness and maintain it in the long term. The collective memory of extreme flood waves and of the measures taken as part of the disaster control, play an important role and memory fades with the years. For that reason France does not make constructions for flood protection that are seldom used, as people forget the meaning of it.

How the responsibility is spread over government, citizens and other parties differs strongly between countries (Gebre & Bruen, 2002). In Germany, Switzerland and the Netherlands society requires absolute protection from floods and other disasters. The preparedness in these countries is low making the society socio-psychologically vulnerable. The governments in these countries want to change this attitude and make a transition to a more risk-aware and well-prepared society. In countries such as the United Kingdom and the United States the focus is already placed on personal responsibility and preparedness. Thus, technologically advanced countries are not necessarily doing so well on awareness raising compared to less developed countries such as Bangladesh, for example, where the people demonstrate great resilience and skill in coping with floods. All countries have much to learn from each other (Gebre & Bruen, 2002).

6.3.4 Living with (uncontrolled) floods

Natural river valleys have river banks gradually sloping upwards. In times of flooding, water depths mostly remain restricted to less than 1 meter. This situation is quite different from that of low-lying polders protected by levees. When a flood of such polders occurs, the water depth may rise to several meters causing serious danger of drowning. The upstream part of the Meuse river in the Netherlands is a natural valley. In 1993 and 1995 the river banks were flooded. As human life is not endangered people living on these banks generally do not feel unsafe (Vlieger et al., 1998). Before the flood in 1993 most of the people were not aware of the flood risk. After the flood the majority of the people felt tensed and sick and one third of the people were not able to go to work. One quarter of the people who felt sick needed more than five weeks to recover. However, most of the people living along the river Meuse are strongly devoted to their resident area and will not move notwithstanding their experience with flooding (Slootweg & Schooten, 2002). Low expectations of selling prices of houses after the flood also played a role.

In the United Kingdom research was done on the health effects resulting from a flood. The health effects are often very marked, ranging from premature death to general feelings of ill-health (FHRC, 1999). Hazards such as floods can be regarded as potentially multi-strike stressors made up of a number of different components. Along with the disruption to households from the flood event itself, factors associated with the disruption from recovering after the flood can also be identified as affecting people's health. The main impacts of recovery disruption resulted from the following: having to leave home; lack of practical and emotional support; lack of advice on what to do; problems dealing with insurers and builders; stress from living in damp and damaged properties; and increased financial worries (FHRC, 1999).

The loss of memorabilia and sentimental possessions are often the cause of great distress to many flood victims. This loss is seen to undermine people's individual sense of self identity and place identity. People develop a sense of self, based on the places in which they live. The home is often conceived as an emotional sanctuary and a haven from the outside world. A personal style created over the course of years needs to be re-created in weeks or months and the choice of furniture and fittings is often based on a fear of future flooding rather than the preferred style. Some flood victims in the United Kingdom described a sense of violation and invasion in the home, which was no longer a secure place to live (FHRC, 1999).

Women generally suffer more from the disruption to the home and often have to deal more directly with the recovery from the flood, in addition to bearing the main responsibility for their family's health care. For foreign women the impacts of recovery are exacerbated when these women lack the ability to speak and understand the national language and are confined to their damp homes for much of the day (FHRC, 1999).

After the repairs have been completed and the painful memories have faded, stress is relieved, and people can resume living their normal lives.

6.3.5   Living with inundation

People find man-made disasters more threatening than natural disasters. Man-made disasters also remain longer in the collective memory of people (Valk et al., 2003). Half of the people living along the rivers in the Netherlands think that floods have natural causes. The other half believes that floods are caused by men (Vlieger, 1998). When calamity polders are inundated, flooding occurs on purpose and thus flooding is completely man-made. That affects the perception and acceptance of the risk of the people living in the calamity polder:
- they find the flood compulsory;
- they might consider the division of risks and gains unfair;
- they might have doubts about the beneficial effects on society;
- they might develop mistrust of government and experts.

These effects enlarge the negative emotions and feelings that people have when water levels rise and flooding threatens. Moreover, inundation will probably get a lot of attention in the media. Altogether, people living in a calamity polder will consider inundation more threatening than an uncontrolled flood. This will be detrimental to the flood risk acceptance.

Assigning a calamity polder to protect other polders from flooding means that the frequency of flooding of that polder becomes larger than that of the other polders. Already rumours on the possibility of assigning a calamity polder get people in that polder worked up; feelings run high. This and later on the discussion with the decision makers generate feelings of suspense, uneasiness and fear among the people. This is aggravated when people believe that evaluation of an eventual assignment is not done carefully enough and think that not all other alternatives to safeguard polders in the river basin area have been sufficiently elaborated. People in a potential calamity polder also get upset when it is not clear to them what the country will do in return for the sacrifice that is asked of them (Baan & Rabou, 2002).

Actually assigning a calamity polder may start social and economic processes that influence the regional development of the area. Some people will move away from the polder due to the risk and the changes made in the landscape. The dwellers fear that this will cause social disruption to the mostly rural and coherent village communities. The native population particularly dislike that and consider the social disruption as the greatest loss (Werff, 2000).

People also fear that ground and property values will fall, and that the economic development in the area will be retarded due to less willingness to invest in the area. The uniqueness and splendour of the landscape might be spoiled as well as the rural, characteristic nature of the area (Slootweg & Schooten, 2002).

Actually inundating the polder will cause damages which are both tangible and intangible. People living in the polder must be evacuated for weeks and fear the inconvenience, the tension and the social disruption, when this should happen. People also fear environmental effects due to (polluted) river sludge settling in the polder area and other pollution caused by flooding (Baan & Rabou, 2002).

How the calamity polder will develop in reality after assignment is uncertain. Based on the current flood protection policy the frequency of flooding of a calamity polder in the Netherlands is only once every 1250 years. With a frequency that low, the government thinks that the current spatial and economic development will go on normally. How the citizens and businesses will actually behave in the future depends on the perception they have of the consequences of the assignment and the spatial planning.

6.3.6   Compensation of damage and insurance

As the flooding of calamity polders is deliberate in order to safeguard other more vulnerable polders, damage must be compensated for. At present the public's experience with damage compensation in the Netherlands is bad. It is stated that with inundation generous settlements of damages are needed (Baan & Rabou, 2002; Slootweg & Schooten, 2002; Baan, 2003). The damage compensation should include not only economic damage, but also an allowance for intangibles like the inconvenience caused by the flooding and the impacts it has on the rural communities. Generous settlements of compensation will contribute to gaining a social basis for the assignment of calamity polders and to a change of the NIMBY effect (Not-In-My-Back-Yard) to an AIMBY one (Allowed-In-My-Back-Yard).

Security that flood damage will be compensated for decreases the threat of flooding. It turns out that dwellers who are insured cope better with the impacts of a disaster. People who have to ask for financial assistance from the government after a disaster experience feelings of stress, dependency and inferiority (Valk et al., 2003). Moreover, in the Netherlands the average citizen finds it unfair that he or she should bear the consequences of events that are not the result or fault of his or her actions (RVZ, 2001).

The most efficient way to cope with a flood's destructive forces is based on co-operation between the people and the government plus the insurance industry (Kron & Thumerer, 2002). Only combined efforts that include initiative, activities, awareness, preparedness, appropriate responses to a flood, etc. by these three groups allow minimisation of the total costs and prevent and mitigate the impacts of floods.

The demand for flood insurance is growing (Kron & Thumerer, 2002). Flood insurance must be made adequate and efficient (Yeo, 2002). Insurers need to adopt incentives for risk reducing behaviour, so that efforts to mitigate damages are encouraged, not undermined. Otherwise the potential damage will increase in future.

Compensation for flood damage is not regulated in the Netherlands neither can people insure themselves against flood risk. Other countries have settled damage compensation (e.g., France after a formal admission of a natural disaster, and Hungary in the case of inundation). Flood insurance is available in, for example, the United Kingdom, Australia and the United States (under restrictions) (Gebre & Bruen, 2002).

6.4   DEVELOPMENT OF A FLOOD RISK MANAGEMENT STRATEGY

The flood protection level along rivers in the Netherlands is very high and uniform. Flood protection is directed almost completely at reduction of flood frequency. When necessary in order to maintain the protection level, structural adjustments are made. Non-structural methods of flow regulation (catchments land use management), vulnerability reduction measures (land zoning, changing flood plain cropping patterns, building design) or emergency measures (sand bags, flood logs) or loss pooling schemes (insurance) receive almost no attention. When drafting a flood risk management strategy both structural and non-structural measures should be addressed. The strategy should aim at making communities in the river basin less vulnerable to floods and preferably should be anticipative and not responsive (Stefanovic, 2003).

We must know and understand how the systems work, what the causes of the (increasing) risks are, and how we can manage the developments. We should not want to shift off problems and losses (to future generations); we should aim to eliminate losses. We also have to realise that the answers we get depend on the questions we put (Stefanovic, 2003). Our paradigms, attitudes and beliefs influence: (i) the scientific questions we put; (ii) the choices we make related to the approach and methods used when searching for solutions; and (iii) how we interpret and analyse the data and the results of analyses. Integral and sustainable solutions can be found only when qualitative research on human values and perceptions and how we can change them forms an integral part of the total research programme.

Developing a flood risk management strategy should be done in close co-operation with the citizens and other parties, and especially with people and businesses that may be affected. They should be involved from the start of the policy process (problem formulation) to the end of it (evaluation of possible measures and decision making). Communicating with all parties involved should get the utmost attention. The information must be given in a language people understand easily, uncertainties must be made clear, and the feelings of the people involved must be taken into account. In addition to the questions and choices mentioned previously, in defining a flood risk management strategy attention should be given to disaster control. This comprises, among others things, monitoring, early warning systems, alarm and evacuation plans, as well as relief services and speedy reconstruction of damaged infrastructure (Kron & Thumerer, 2002). While drafting alarm and evacuation plans and plans to organise relief services (practical and emotional support), special attention must be given to vulnerable parties (FHRC, 1999; Slootweg & Schooten, 2002).

REFERENCES

Arvai, J.L. 2003. Using Risk Communication to Disclose the Outcome of a Participatory Decision-Making Process: Effects of the Perceived Acceptability of Risk-Policy Decisions. *Risk Analysis*, Vol. 23, No. 2, pp. 281–290.

Baan, P.J.A. 2003. Coping rationally with flood risks? (in Dutch: Nuchter omgaan met overstromingsrisico's?). *H_2O*, (36) 2003, nr. 23, pp. 41–43.

Baan, P.J.A. & Klijn, F. 2003. Safety approach for flooding: improving preparedness? (in Dutch: Veiligheidsbenadering bij overstromingen: naar meer zelfredzaamheid?) *H_2O* (36) 2003, nr. 18, pp. 17–18.

Baan, P.J.A. & Rabou, J.P. 2002. Feelings with assignment and use of calamity polders and side rivers (in Dutch: *Gevoelens bij aanwijzen en gebruik van retentiegebieden en groene rivieren*). WL | Delft Hydraulics, report Q2975.25. Delft, the Netherlands, March 2002.

Barnes, P. 2002. Approaches to community safety: risk perception and social meaning. *Australian Journal of Emergency Management*, autumn 2002, pp. 15–23.

Dijkman, J. et al. 2003. Explanation of expert judgement for calamity polders (In Dutch: *Toelichting aanvullend deskundigenoordeel noodoverloopgebieden.*) WL | Delft Hydraulics, report Q3570. Delft, the Netherlands. August 2003

Flood Hazard Research Centre (FHRC) 1999. The health effects of floods. The Easter 1998 Floods in England. *Flood Hazard Research Centre Article Series*, No. 3/99

Flinterman, M.H., Glasius A.T.F. & Konijnenburg, P.G. van 2003. The perception of flood risks (in Dutch: *De perceptie van overstromingsrisico's*). Bouwdienst Rijkswaterstaat. Utrecht, the Netherlands. January 2003

Gebre, F.A. & Bruen, M. 2002. *Worldwide Public Perception of Flood Risk in Urban Areas and Implications for Policy Formulation.* (www.mitch-ec.net/workshop3)

Kron, W. & Thumerer, T. 2002. *Water-related disasters: Loss trends and possible counter-measures from a (re-)insurers viewpoint.* Munich Reinsurance Company, Germany (http://www.mitch-ec.net/workshop3/Papers/paper_thumerer.pdf)

Lyklema, S. 2001. Water management and communication: a study on the social basis of water management in the Netherlands (in Dutch: *Water beheren en communiceren: een studie naar het publieke draagvlak over het waterbeheer in Nederland*). Thesis Wageningen University, the Netherlands

Penning-Rowsell, E. 2003. Implementing flood mitigation and protection: constraints, limitations, power and "reality". In: M. Marchand, K.V. Heynert, H. van der Most & Penning W.E.(eds). *Dealing with flood risk. Proceedings of an interdisciplinary seminar on the regional implications of modern flood management. Delft Hydraulics Select Series 1/2003.* Delft University Press

Rees, J.A. 2002. *Risk and Integrated Water Management.* Global Water Partnership, TEC Background papers, No. 6. Eleanders Novum, Sweden, August 2002

Rijksinstituut voor Volksgezondheid en Milieu (RIVM, 2003). Coping rationally with risks (in Dutch: *Nuchter omgaan met risico's*). Milieu-en natuurplanbureau RIVM, report 25170104/2003 (www.rivm.nl/bibliotheek/rapporten/251701047.html)

Raad voor de Volksgezondheid & Zorg (RVZ, 2001). Prediction, prevention and insurance of health risk (in Dutch: *Gezondheidsrisico's voorzien, voorkomen en verzekeren*). Advice to the minister of Health, Well-being and Sports, Zoetermeer, the Netherlands.

Scherer, C.W. & Cho, H. 2003. A Social Network Contagion Theory of Risk Perception. *Risk Analysis*, Vol. 23, No. 2, pp. 261–267

Sjöberg, L. 2000. Factors in Risk Perception. *Risk Analysis*, Vol. 20, No. 1, pp. 1–11

Sjöberg, L. & Drottz-Sjöberg, B.M. 1994. *In Radiation and society: comprehending radiation risk. Proceedings of an International Conference in Paris, 24–28 October 1994.* Vol I, Wenen, International Atomic Energy Agency

Slootweg, R. & Schooten, M. van (2002). Social aspects of calamity polders (in Dutch: *Sociaal-maatschappelijke aspecten van noodoverloopgebieden*). Report SEVS Beleidsadvies voor natuur en leefomgeving. Oegstgeest, the Netherlands, 14th March 2002

Slovic, P., Finucane, M.L., Peters, E. & MacGregor, D.G. 2002. *Risk as Analysis and Risk as Feelings. Some thoughts about Affect, Reason, Risk and Rationality.* Paper presented at the Annual Meeting of the Society for Risk Analysis, New Orleans, Louisiana, December 10, 2002 (www.decisionresearch.org/pdf/dr502.pdf)

Slovic, P. & Weber, E.U. 2002. *Perception of Risk Posed by Extreme Events*. Paper prepared for discussion at the conference "Risk management strategies in an Uncertain World". Palisades, New York, April 12–13, 2002 (www.ldeo.columbia.edu/res/pi/ CHRR/Roundtable/slovic_wp.pdf)

Stefanovic, I.L. 2003. The Contribution of Philosophy to Hazards Assessment and Decision Making. *Natural Hazards*, Vol. 28, pp. 229–247

Valk, M., Heinen, H., Leineweber, & Otten, J. 2003. Awareness of water in the Netherlands: learning of risk awareness processes in foreign countries (in Dutch: *Waterbewustzijn in Nederland: Leren van risico-bewustwordingsprocessen in het buitenland*). RIKZ/ Bouwdienst/Ergo, Report nr. RIKZ/2003.005, the Hague, the Netherlands

Vlek, C.A.J. 2001. Psychology of risks: each advantage has its risk (in Dutch: Risicopsychologie: elk voordeel heeft zijn risico). *Hypothese, kwartaalblad voor onderzoek en wetenschap*, jaargang 8, no. 31, pp. 12–15

Vlieger, W. de, Lageweg, M.C.C., Vollering, D.C. & Bakker, J. 1998. Study on experiences and feelings with high water levels (In Dutch: *Belevingswaardenonderzoek: Risicobeleving Hoogwater*). Bouwdienst Rijkswaterstaat, the Netherlands. February 1998

Weber, E.U., Blais, A.R. & Betz, N.E. 2002. A Domain-specific Risk-attitude Scale: Measuring Risk Perceptions and Risk Behaviors. *Journal of Behavioral Decision Making*, Vol. 15, pp. 263–29

Werff, P.E. van der 2000. *Nature or neighbour in Hell's Angle: Stakeholders response to future flood management plan for the Rhine River*. IVM report D-00/10, SIRCH Working paper 9, Free University Amsterdam, December 2000

Yeo, S.W. 2002. Flooding in Australia: A Review of Events in 1998. *Natural Hazards*, Vol. 25, pp. 177–191

7

The Role of Private Insurance Companies in Managing Flood Risks in the UK and Europe

David Crichton
Benfield Hazard Research Centre, London, UK

ABSTRACT: This paper outlines the different approaches to state and private flood insurance in Europe and around the world, with particular reference to the United Kingdom. The UK is unique in the world in that private insurers agreed in 1961 to guarantee to include cheap flood insurance as part of the standard household package policy for all households, regardless of the hazard. This means that around 95% of households in the UK have flood insurance.

This high penetration of cover has justified the investment of considerable resources by insurance companies in UK flood mapping and modelling for both coastal and river floods and indeed some insurers arguably have better flood maps than the government. Insurers have also pooled their detailed claims data and have funded research to identify ways in which buildings could be made more resilient.

The author's "insurance template" indicates what levels of flood hazard are acceptable to insurers and has been widely adopted by planning authorities in Scotland. Unfortunately demand for housing in England and Wales means that 27% of new homes are now being built in flood hazard areas.

As a result, UK insurers have been forced to review the 1961 guarantee. In 2001 the price cap was removed, and at the end of 2002, the guarantee was removed altogether for areas where the flood hazard exceeds 1 in 75 years.

7.1   PRIVATE FLOOD INSURANCE

There is a wide variation around the world in the use of private insurance to cover flood events (van Schoubroeck, 1999). There is also evidence that as climate change increases the frequency and severity of floods around the world (Crichton, D., 2001) some governments around the world are becoming concerned about the apparently growing number of severe flood events, especially in those countries where flood victims are not generally protected by insurance.

In those countries where there is private insurance for residential properties, it can be categorised into two types, the "option" system and the "bundle" system (Crichton, D., 2002a).

7.1.1   The option system

Under this system, insurers agree to extend their policy to include flood on payment of an additional premium. This system can be found in Australia (Queensland and Northern Territories only), Belgium, Canada (commercial properties only), Germany, and Italy.

There are a number of problems with this system (Crichton, D., 2002b). The biggest is adverse selection. Insurers tend to select against customers by only making the cover available in areas they consider to be safe, while customers select against insurers by only buying it in areas they deem to be risky. The result is that cover when it is available at all, is expensive, and has very low market penetration. It is unlikely to be sustainable.

7.1.2   The bundle system

In this system, cover for flood is only available if it is "bundled" with other perils, such as storm, theft, earthquake, etc. This system is used in:

Israel, Japan, Portugal, and the UK.

Insurers have the freedom to charge differential rates, but excessive rate increases can be mitigated because the risk is not only spread over time, but across perils, and across postcodes. People living in areas safe from flood still have to buy flood cover, if they want to get earthquake cover, for example, and vice versa.

This system seems to produce universally much higher market penetration, ranging from 40% in Japan to 95% in the UK.

While the opportunities for adverse selection by customers are minimised, there could still be some "cherrypicking" by insurers.

For a useful discussion on different types of flood insurance systems see the Focus Report published by Swiss Re (Menzinger, I., and Brauner, C., 2002).

In the UK there is a practice amongst finance houses that lend money to those wishing to buy a house, to insist that the house be insured in order to protect their collateral. These finance houses will usually arrange block policies with a preferred insurer who then has to accept any property his principal decides to lend money on. Recent changes in the rules means that the individual does not have to insure under the block policy and some insurers are actively encouraging such borrowers to move their insurance to them. They only offer this encouragement of course, where the borrowers live in a safe area, and ultimately there could be a vicious circle situation for the finance house.

7.2 THE ROLE OF THE STATE IN THE COMPENSATION OF FLOOD VICTIMS

Apart from the USA, which has a rather different system, there are three categories of state involvement in compensation for flood victims, and these are outlined below. First, however a general point needs to be made; there seem to be very few examples of a State deciding to arrange for reinsurance cover from foreign reinsurers, all the costs are kept within the boundaries of the country. This could be dangerous if there is a major catastrophe or series of catastrophes because the State may find itself in financial difficulties or at the very least find it difficult to attract capital investment in the future without some element of compulsion, such as a levy or insurance tax (Crichton, D., 2002c).

Reinsurance offers a very cost-effective way to spread the risks across the economies of other countries so that if one country is hit by disaster, other countries automatically step in with support.

No state compensation for citizens (*although there may be grants for infrastructure*) Argentina, Germany, Israel, Japan, Portugal, UK. This causes severe problems for those where insurance is not available or is not affordable. Such people are often the hardest hit when a flood strikes, and the resulting stress and other health consequences could seriously affect the economy.

7.2.1 Procedures to provide compensation in hardship cases

7.2.1.1 *Australia, Canada, and China*
This is a pragmatic solution, but not necessarily the best. The State may not be well geared up to assess how much compensation to pay or to administer it efficiently, whereas insurers have the systems in place to pay out fair levels of compensation. Also some citizens might regard such compensation as "charity" and be too proud to accept it.

7.2.2 Compensation only by decree after the event

7.2.2.1 *Belgium, France, Italy, and Spain*
This can be dependent on the whim of a politician who could be influenced by many other issues, not all entirely objective. The main effect is that while such a scheme can be very expensive for the State, the citizen is never sure in advance if he will be protected, and may be reluctant to invest in his property if he thinks it may be flooded.

7.2.3 Conclusion

If insurers want to earn a "seat at the table" when it comes to Government policy decisions about flood risk reduction, they need to be prepared to provide affordable flood cover for all but the most extreme risks. At the same time, if the State wants private insurance to flourish, it needs to make a conscious decision *not* to provide compensation to its citizens but to concentrate on renewing infrastructure and making sure that insurance is widely available and affordable. In other words, a partnership approach between the insurance industry and the State is essential. For a risk to be insurable, there are a number of requirements. The Mnemonic "BASIC MUD" sums these up (for details, see Appendix 2).

7.3   ADVERSE SELECTION AND "CHERRYPICKING"

Where the customer knows more about his risk than the insurer, or where the insurer, through ignorance, incompetence, regulation, or market forces has failed to recognise the extent of the risk with adequate premium levels, the customer can select against the insurer. With flood insurance, only the customer in the flood risk areas will want to insure for flood, for example. The corollary of adverse selection is often called "cherrypicking". This is where an insurance company decides to only provide insurance for the most profitable, relatively claims free types of risks. Some insurers, for example, will only offer motor insurance to mature drivers with family cars and a good claims record.

While some larger insurers will use the term "cherrypicking" in a derogatory sense, it is a perfectly valid strategy in a free market, especially for smaller insurers trying to build up a profitable account. It is a less viable strategy for the large insurer trying to maintain a large book of business, and in such cases appropriate pricing strategies are not enough, the insurer will need to offer assistance and incentives to policyholders to manage and reduce their risks wherever possible. In particular, for flood and storm hazards, resilient reinstatement is important.

7.4   WORLD WATER FORUM 3

In a briefing paper for this meeting, it was stated that:

"On the World Water Forum 3 in a special session on 'Urban Floods Risk Mitigation', it was emphasised that there is an urgent need to share relevant information and experiences between nations and actors and to initiate joint research activities, in order to reduce and better control the catastrophically impacts of floods world-wide".

"The session concluded with recommendations for further action in three cross-sectoral areas: planning policies and instruments, technological and structural options focussing on adaptation strategies to climate variability and change in the building environment (such as improving the flood resilience of the built environment), and flood risk assessment and risk sharing and spreading."

The insurance industry is well placed to provide advice on adaptation strategies, risk sharing and sharing of data on flood issues, thanks to its extensive experience in the UK market.

7.5   DIFFERENT TYPES OF FLOOD INSURANCE

Flood affects different classes of insurance in different ways, and insurers' strategies in the UK are currently changing dramatically (Clark, M., Priest, S. J., Treby, E. J., Crichton, D., 2002). Here is a summary of the main classes.

7.5.1   Household property insurance

This is usually written as a package of covers for owners or occupiers of domestic residential properties. In Britain, flood is included in the package and therefore anyone wanting fire, theft, earthquake etc insurance must accept and pay for flood cover too. This reduces adverse selection.

7.5.2   Commercial property insurance

In Britain, flood cover may be included in package policies issued to small businesses such as shops, hotels, offices, surgeries etc. However for larger commercial premises, it is normal to have separate policies for fire, theft, and liability etc insurances. The commercial fire policy can be extended to include "wet perils" such as storm, tempest, burst pipes etc and flood. Flood cover is usually only available if storm and tempest cover is included.

So, for large commercial policies, flood cover is an "optional extra". This gives problems with definitions, and also encourages adverse selection. However it does mean that the underwriter can consider flood as a separate peril and underwrite or decline that part of the risk on its merits.

7.5.3   Business interruption insurance

"Just in time" stock control means that industry well outside flood hazard areas may suffer from disruption of supplies or transport problems.

7.5.4   Contractors all risks insurance

Construction sites can be especially vulnerable due to disruption of drainage systems and spread of contamination to third party sites.

7.5.5   Liability insurance

Liability underwriters do not usually consider flood, but there are many reasons why they should think more about such risks, as there might be liability to employees or the public arising from injury or disease caused by flood waters.

7.5.6   Professional indemnity insurance

Architects, and other building professionals, could face expensive litigation in the future if they are involved in developments in known flood hazard areas.

7.5.7   Motor insurance

Not only damage to parked vehicles, but damage to vehicles being driven on flooded roads. For example, water can enter the engine through the exhaust system, or the vehicle can be swept away even by quite shallow water if the water is moving quickly. Most flood fatalities in Britain in recent years have involved motor vehicles.

7.5.8   Life and pensions assurance

Flood fatalities have fortunately been rare since 1953 but much more research is needed into the effects of illness, over exertion, and psychological impacts of flooding on life expectancy.

7.6   CLAIMS ISSUES

Insurers are aware that flood or storm damage can lead to exaggerated claims and that there are many opportunities for fraud. If for reasons of solidarity, there is a desire to compensate flood victims, it is more efficient and fairer if this is done by insurers, who have the expertise to reduce fraudulent claims.

7.6.1   Encouraging private flood insurance in The Netherlands

There are a number of points to be considered.

7.6.2   State compensation

Private flood insurance has little chance of success if the State continues to provide even limited compensation to flood victims. However State payments need not be withdrawn altogether:
- it could be limited to, say €1,000 per household, and insurers could apply a €1,000 excess, so they would be picking up losses in excess of that amount;

- instead of compensation after an event, regular payments could be paid in the form of a subsidy on insurance premiums for low income and elderly people as part of the social welfare scheme. In this way the State is targeting its resources to those who need it most, and these costs are known in advance.

Table 7.1. Standard protection of different types of housing (expressed in return period).

Type of housing	Return period (years)
Sheltered housing, and homes for the disabled and elderly	1,000
Children's homes, boarding schools, hotels, hostels	750
Basement flats	750
Bungalows without escape skylights	500
Ground floor flats	500
"Flashy" catchments (little or no flood warning available)	500
Bungalows with escape skylights	300
Caravans for seasonal occupancy only, provided adequate warning notices and evacuation systems are in place	50
All other residential property	200

It should be noted that flooding gives ample scope for fraudulent or criminal activity and insurers are well experienced in dealing with such incidents.

Insurance has very efficient methods of spreading risk globally through the reinsurance industry so that after a major event, premiums do not necessarily have to increase.

7.6.3   Land use planning

Insurers would expect to see tight controls on land use for new housing developments to ensure that no new buildings are erected in areas of high flood hazard. They will take into account the standard of protection offered by flood defences however, and where the standard of protection is higher than 1 in 200 years, flood cover should be possible. Initially until the insurance industry has gained confidence, they may insist on a standard of protection of 1 in 1,000 years.

Local planning authorities should ideally be legally and democratically accountable for building and maintenance of flood defences and watercourses, to give them an incentive to avoid increasing exposure to flood hazards, by allowing new building in hazardous areas.

In Scotland, local planning authorities now follow some or all of the "insurance template" which sets out levels of risk acceptable for flood insurance at normal terms (Crichton, D., 2003c). It is not intended to dictate to planners, what their policy should be, rather it is to inform them that if they allow development where the risk exceeds the template, there may be difficulties in arranging insurance on the property.

Extract from the residential property section of the "insurance template" (Crichton, D., 1998) ©*Copyright* Crichton, 1998.

Return period up to the year 2050 in each case, taking climate change into account.

7.7   THE "ONTARIO SOLUTION"

Following Hurricane Hazel in 1964, the province of Ontario, in Canada, decided that certain areas, equating to the 1 in 200 or 1 in 250 year flood hazard should no longer have any flood defence expenditure other than on evacuation routes, and no new properties should be permitted there (Brick, J., and Goldt, R., 2001). Instead the community has funds to buy up properties in such areas and no one is allowed to sell such properties other than to the community. The result is many areas of parkland next to rivers, and attractive forest walks.

7.8   DROUGHT DAMAGE TO FLOOD DEFENCES

The breach of a defence in Amsterdam in summer 2003, should remind the authorities of the fragility of peat based defences in drought conditions. PS InSAR is a technique involving satellite data, which can give early warning of landslip or subsidence. Insurers would wish to see widespread use of this cheap and effective technique in order to monitor the stability of flood defences. Already, Italy has installed nearly 700 "corner reflectors" which mark critical areas for monitoring and which can be easily picked up by the satellites.

7.9   RISK ASSESSMENT

The flood models and maps produced by reinsurance brokers such as Benfield Group Ltd and other leading brokers like AON, Guy Carpenter, and Willis, or major reinsurers like Swiss Re, are becoming increasingly sophisticated. The State could commission studies involving its own academic research institutes working with insurance modellers to quantify the risks, and identify areas for flood defence spending, or "managed realignment", based on the potential flood losses from building damage and business interruption.

7.10   CATASTROPHE RISK ACCUMULATION

Insurers will be concerned about the maximum possible loss arising from a catastrophe situation such as an extreme flood event. One solution is to "compartmentalise" the country, rather like waterproof doors in a ship, or fire doors in a building, so that if a catastrophe occurs it can be contained within one area of the country.

7.11   LIMITED FLOOD COVER

There may be a temptation on the part of insurers to offer cover just for sewage backup flooding. This should be resisted, owing to the arguments that take place after every flood about the definition of flood, and whether the policy should operate. There is also the risk of collusion with loss adjusters. A further important point is that if even limited flood cover is offered in high hazard areas, this could encourage people to continue to build in such areas, resulting in maladaptation (Crichton, D., 2003b).

7.12   OPTION V BUNDLE

An option system is unlikely to succeed, based on the experiences in other countries (see above). Discussions should be held with insurers and mortgage lenders to assess their willingness to agree to a market wide approach to offer flood as part of a bundled system for packaged household and small business policies. Ultimately, the mortgage lending industry could insist on flood cover as a condition of lending to property purchasers.

7.13   FLOOD RESILIENCE

Insurance data shows that many parts of the fabric of a building are particularly vulnerable to flood damage (Black, A., and Evans, S., 1999). Building regulations should be reviewed to ensure that new buildings are constructed in such a way that they will suffer minimal damage from flooding. New regulations should, in due course be made retrospective so that reinstatement has to be to the new standards.

7.14   COMMENTS

Insurers cannot be altruistic when it comes to offering cover against natural disasters; they have shareholders to answer to. Despite this, the government and the media often seem to expect that insurers should cover everything and everyone, and the more cover they give, the more is demanded. Insurance can benefit society by helping with recovery after a disaster, and in carrying out research into the hazard (Crichton, D., and Mounsey, C., 1997). However, as has been seen in the UK, especially after the autumn floods of 2000, and 2002, there is a danger that society may take the availability and affordability of insurance for granted and see it as a social service rather than a commercial enterprise (Entec Ltd and JBA Ltd, 2000).

As insurers rethink their strategies in the light of climate change and other pressures, so society may have to rethink the role of insurance.

REFERENCES

Black, A., and Evans, S., 1999 *"Flood damage in the UK: New insights for the insurance industry."* University of Dundee. ISBN 0 903674 37 8. Dundee, Scotland.

Brick, J., and Goldt, R., 2001 *"City of London Flood Plain Management"*. Upper Thames River Conservation Authority, London, Ontario, Canada.

Clark, M., Priest, S. J., Treby, E. J., and Crichton, D., 2002 *"Insurance and UK Floods: a strategic reassessment."* A Research Report for TSUNAMI (now the "Risk Group"). University of Southampton, Southampton.

Crichton, D., and Mounsey, C., 1997 How the Insurance Industry is using its flood research. *Proceedings of the 32nd MAFF Annual Conference of Flood and Coastal Engineers*, Keele, England.

Crichton, D., 1998 Flood Appraisal Groups, NPPG 7, and Insurance, in Faichney, D., and Cranston, M., (eds), *Proceedings of the "Flood Issues in Scotland" seminar* held in Perth in December 1998. Scottish Environment Protection Agency, Stirling, Scotland, pp 37–40.

Crichton, D., 2001 *"The Implications of Climate Change for the Insurance Industry."* (ISBN 1-903852-00-5), Building Research Establishment, Watford, England. Available from http://www.brebookshop.com/ stock reference 42524.

Crichton, D., 2002a *"UK and Global Insurance Responses to Flood Hazard."* Water International Vol 27, 1. pp 119–131. Illinois, USA.

Crichton, D., 2002b Residential Flood Insurance: Lessons from Europe. In Smith, D. I., and Handmer, J., (eds) *Residential Flood Insurance*, 41–58, and 293–295. Water Research Foundation of Australia, Canberra.

Crichton, D., 2002c *"The 'flood tax' – is the Government out of its depth?"* Town & Country Planning, 2002, vol 72, March, pp 66–68. London.

Crichton, D., 2003a *"Best practices in risk management in the context of climate hazards"*. United Nations Framework Convention on Climate Change workshop on "Insurance related actions to address the specific needs and concerns of developing country Parties arising from the adverse effects of climate change" Rheinhotel Dreesen, Bonn, Germany. See: http://unfccc.int/sessions/workshop/140503/present. html See also: http://unfccc.int/sessions/workshop/140503/ documents/crichtonrisk.pdf

Crichton, D., 2003b *"Insurance and Maladaptation to climate change"*. United Nations Framework Convention on Climate Change workshop on "Insurance and risk assessment in the context of climate change and extreme weather events" Rheinhotel Dreesen, Bonn, Germany. See: http://unfccc.int/sessions/workshop/120503/ present. html See also: http://unfccc.int/sessions/ workshop/120503/documents/ crichtonadapt.pdf

Crichton, D., 2003c *"Flood risk and insurance in England and Wales: are there lessons to be learnt from Scotland?"* Technical Paper Number 1, Benfield Hazard Research Centre, University College London. Available for free downloading from www.benfieldhrc.org/SiteRoot/activities/ tech_papers/flood_report.pdf

Entec Ltd and JBA Ltd, with contributions from Crichton, D., and Salt, J., 2000 *"Inland Flooding Risks. General Insurance Research Report No 10"* Association of British Insurers (ABI), London.

Menzinger, I., and Brauner, C., (Swiss Re), 2002 "Floods are insurable" Swiss Re Focus Report, Zurich, 2002.

van Schoubroeck, Caroline, Center for Risk and Insurance Studies, KU Leuven University, Belgium, 1999 *"Flood Insurance Systems in Selected European Countries"*. Published in Proceedings of the Euro Conference on Global Change and Catastrophe Risk Management, Laxenburg, Austria, 6–9 June, 1999.

ANNEX 1

Recent history of flood insurance in Britain

Britain is unique in that in 1961, all insurers agreed that not only would they include flood insurance cover in every household insurance package policy, they would not charge more than £0.50 premium per £100 sum insured for such cover, regardless of the risk. In effect, owners of property in safe areas were subsidising the premium of those in hazardous areas. This was not really a gesture of solidarity on the part of those in safe areas, because almost none of the public was aware of the "guarantee" and in any case the subsidy made very little difference to their premium (less than 4%).

There was therefore a very high market penetration of flood insurance (95%) because:

1. flood cover was "bundled" into the package policy so there was no option but to take the cover. This prevented adverse selection and ensured a spread of risk;
2. mortgage lenders insisted that the full policy package be insured before they would lend money to buy the property;
3. government refused to pay any compensation to flood victims.

From the insurers' point of view the system worked well initially because at the time there were relatively few flood events, and in any case there were no high-resolution flood maps which underwriters could use. Also the coastal defences built after the 1953 floods were still relatively new and there was a high level of confidence in them.

Following the coastal floods in 1990 in Towyn, North Wales, insurers started to become concerned about the poor condition of sea defences in England and Wales, and the numbers of properties which had been built in hazardous areas since 1961. Many insurers were already using geographic information computer systems (GIS) for marketing purposes, and some started to extend their use into flood risk mapping and modelling. The market collectively started to fund research through the Association of British Insurers (ABI) into the condition of sea defences around England and Wales. They were shocked to find that due to lack of maintenance, more than 1,200 defences were likely to fail in a 1 in 50 year coastal storm.

The 1993 floods in Scotland (Perth) and North Wales (Llandudno), were followed by the 1994 floods in Scotland (Strathclyde), the 1998 floods in the Midlands of England, and the 2000 floods in Edinburgh in April and most of England in the Autumn. It was clear that after several decades from 1961 to 1990 with very few flood events, flooding was now becoming a real issue.

This was due to increases in:

– hazard: Climate change plus a deterioration in flood defences due to their age and lack of maintenance;
– exposure: Many more properties were being built in flood hazard areas due to demand for housing;
– vulnerability: Inadequate resilience in building standards combined with greater use of lightweight building materials and materials vulnerable to flood damage such as composite boards.

There was little that insurers could do about this, and they called on government to
1. invest more in flood defences;
2. control building in flood hazard areas;
3. make building standards more resilient.

These requests were virtually ignored by government in England and Wales, but after Scotland won a measure of independence in 1999, the new Scottish Parliament, the Scottish Executive, and Scottish local government authorities did start to implement these measures.

Following the publication of revised draft planning guidelines for England in 2001, which gave planners the authority to continue to allow new building in flood hazard areas, insurers had had enough. The implied partnership with government had broken down and the price cap was removed. Government were given until the end of 2002 to revise the planning guidelines and take other measures to reduce flood risk or the guarantee would be removed. Government did not respond and so at the end of 2002 the guarantee was duly removed, and replaced with a limited guarantee until 2007. Mortgage lenders agreed that they would not provide loans for properties that did not have flood insurance, and many householders were left in limbo as their policies were cancelled and their properties lost value because they could not be sold.

ANNEX 2

Insurability Issues for flood: a global view

How insurance works
For a risk to be insurable, there are a number of requirements. The Mnemonic "BASIC MUD" (Crichton, D., 2002b) sums these up:
– **B** Big enough "book" of business, i.e. a large enough collection of risks for a statistical spread.

–   **A** Adverse selection minimised through good knowledge of each risk.

–   **S** Sustainable over a number of years for various future scenarios so the risks can be spread over time and reserves built up. Because liability claims can take a number of years to settle, it is possible for casualty insurance to be appear to be profitable in the short term, or as long as the account is growing. If it is not priced properly, however, as soon as the account starts to reduce in size, current income is no longer enough to cover increases in costs of old claims.

–   **I** Information readily available from reliable sources about hazard, vulnerability, exposure and claims triggers. Legislation under the Aarhus Convention will make such information more easily obtainable in the future as the Convention requires disclosure of environmental information to the public.

–   **C** Consistent with existing insurance practices, systems, customs and law.

–   **M** Moral and political hazard low and manageable. For example the country must comply fully with the Harare Declaration, which establishes standards for human and property rights and democratic systems.

–   **U** Uncertainty about the potential loss. At least one of the following must be uncertain:
    –   Will it occur? (for example, property or casualty insurance)
    –   When will it occur? (for example life insurance) – or
    –   How much will it cost? (for example "after the event" insurance);

–   **D** Demand for insurance must exist (or have potential to be created) and must be effective, i.e. there must be enough customers prepared to pay the price that insurers need to charge for providing sustainable insurance.

ANNEX 3

Adverse selection and "cherrypicking"

Where the customer knows more about his risk than the insurer, or where the insurer, through ignorance, incompetence, regulation, or market forces has failed to recognise the extent of the risk with adequate premium levels, the customer can select against the insurer. With flood insurance, only the customer in the flood risk areas will want to insure for flood, for example.

The corollary of adverse selection is often called "cherrypicking". This is where an insurance company decides to only provide insurance for the most profitable, relatively claims free types of risks. Some insurers, for example, will only offer motor insurance to mature drivers with family cars and a good claims record.

While some larger insurers will use the term "cherrypicking" in a derogatory sense, it is a perfectly valid strategy in a free market, especially for smaller insurers trying to build up a profitable account. It is a less viable strategy for the large insurer trying to maintain a large book of business, and in such cases appropriate pricing strategies are not enough, the insurer will need to offer assistance and incentives to policyholders to manage and reduce their risks wherever possible. In particular, for flood and storm hazards, resilient reinstatement is important.

The vicious circle

Commodity business such as household and motor insurance is largely price driven and based on statistical analyses. Commercial business such as property or casualty (fire or liability) cover for factories and offices is written on a more individual basis. For either type of business, accurate underwriting is necessary to avoid the vicious circle. The circle works like this:
1. premiums are too low for the risks insured;
2. losses are incurred;
3. across the board increases in premium are applied;
4. "good" risks, for example householders who are careful and have a low propensity to claim, or businesses which are well managed, will find cheaper premiums elsewhere, (or will chose not to insure at all) and will cancel their policies;
5. because the increases have been inadequate for bad risks, more of these will be attracted, or retained, so a higher proportion of the book will be "bad" risks.

ANNEX 4

(The following is based on a proposal presented by the author to the Australian Government and Insurance Industry at a conference in Canberra in 2001. During the conference a number of options were considered and the favoured option was close to the author's proposal. It was subsequently published in 2002 by the Water Research Foundation of Australia in a book of the proceedings titled "Residential Flood Insurance", edited by Dr David Smith and Prof. John Handmer. The author has adapted it as a possible way forward for the Dutch market.)

The following is based on the assumption that the Dutch government wishes to make private sector flood insurance work without any additional cost implications for government. However, there is a case for saying that for those on old age pension, income support or other form of welfare benefit, government should make a contribution by increasing such benefits to pay for insurance premiums.

Also grants could be given for individual flood protection and resilience measures in extreme cases of high hazard or high vulnerability.

This would be much more cost effective for the government than finding money to pay compensation to people after the event, and much more equitable than the present situation where many people are inadequately compensated.

The "Crichton Formula"

The following notes were written in a UK context, and it is not suggested that they be used for The Netherlands. However they give an indication of how a partnership approach could help to make private flood insurance sustainable.

It should be noted that while this includes obligations on the part of local and national government, it does not in itself involve any increase in taxpayer funded expenditure.

Note the restrictions on the flood excess: it is inequitable to have a flat sum excess of, say £10,000, which could cause great difficulty for low-income families. In any case, the traditional purpose of an excess was to encourage the policyholder to act to avoid or mitigate losses, and there is little that can be done to avoid flooding. However, if the policyholder is fit and healthy and receives enough warning it may be possible to move contents to a safer place.

If this is done it is only fair to reduce the excess in such cases, or in any case where the policyholder is physically unable to move property.

Part one – insurers' obligations

1. *General*
1.1 All insurers which provide property loss or damage cover under household and small business package policies (buildings and contents, including alternative accommodation and business interruption) will include cover for water damage caused by flood, storm or burst pipes, for all such policies. Those insurers will guarantee not to refuse cover on the grounds of flood risk for areas within the boundaries of those local authorities which accept the obligations outlined in Part two of this formula.

1.2 For those local authority areas, the flood excess will not exceed 10% of the sum insured, but there will be no premium limit. Cover will not be refused on the grounds of flood risk unless:
   – the property owner has refused the offer of a flood alleviation scheme;
   – the property has suffered flood damage three or more times in the previous five years and no additional flood protection measures have been put in place since that time.

1.3 For those local authority areas, insurers agree not to seek to recover flood claims costs from the local authority unless there is evidence of gross negligence on the part of the Council, or failure to ensure that all watercourses within the Council boundaries are properly maintained.

2. *For buildings insurance*

2.1 In the event of a flood claim, the maximum amount payable will not exceed the market value of the property immediately prior to the flood. (This is to cater for properties sold cheaply due to a history of flooding in the area.)

2.2 In order to avoid blight, insurers will maintain cover on existing property after it is sold, subject always to the terms of this formula.

2.3 In the event of a flood claim, if the insurer considers future flood claims to be inevitable, it can reserve the right to:
    – pay the market value or outstanding mortgage (whichever is greater) for the property in return for the title deeds. The insurer is then free to demolish or otherwise dispose of the property; OR
    – cancel the policy unless the policyholder agrees to install flood protection measures to the satisfaction of the insurer at his or her own expense.

2.4 The insurer reserves the right to reinstate flood-damaged property in such a way that it will be less vulnerable to future flood events. If betterment is involved the policyholder will be liable for any additional cost (this would not apply where resilient reinstatement is a legal requirement). However the insurer will provide a low interest loan for this additional cost in return for a long-term agreement covering the repayment period. During the repayment period, the insurer will undertake to continue to invite renewal of the insurance at a premium rate not exceeding 20% more than the rate for the previous year, and will continue to invite renewal on this basis even if the property is sold. If the policyholder is not prepared to pay the additional cost of betterment, the insurer is released from any obligation to renew the policy.

3. *For contents insurance*

3.1 Any flood excess will be waived or reduced as appropriate if the policyholder can demonstrate that reasonable precautions were taken to move contents to a place of safety or otherwise reduce the loss, taking into account all the circumstances, including, but not limited to:
    – risks to the health of the policyholder or the policyholder's dependants;
    – the policyholder's state of health or fitness;
    – the amount of warning given;
    – ability to readily access a place of safety.

4. *Insurers shall be released from any obligations where:*
    – the property is within a dam break inundation area or managed retreat or realignment area;
    – the policyholder is deemed by the ABI to be unsuitable for insurance, for example due to criminal activities;

- the flood hazard is greater than 75 years for properties constructed or granted planning consent prior to 2003;
- the flood hazard is greater than the 100 year return period for properties granted planning consent after 2003;
- the flood hazard is greater than the insurance template for properties granted planning consent after 2006;
- the flood hazard is greater than the insurance template for any properties after 2010
- the property has been identified in an ABI advisory notice as having an unacceptably; high level of flood risk, for example where the EA have advised against the development;
- the local authority withdraws or is deemed by the ABI to have withdrawn from the obligations outlined in part two of this formula.

Part two – local Government obligations

This formula will only apply to properties within the boundary of those local authorities that have agreed to:
1. Hold at least four meetings a year of flood appraisal groups including representatives from the relevant environment agency, River Basin Management Board, Water Company or Authority and the insurance industry. All decisions relating to strategies for floodplain management including flood defences, planning, drainage impact assessments, flood risk assessments, and significant developments where flood hazard is a possibility, to be referred to the flood appraisal group.

2. Adopt a presumption in their planning strategies against development for cases where the flood hazard exceeds the insurance template.

3. Make available to the insurance industry timely and detailed information about any proposed new residential development which breaches the insurance template so that the ABI can issue an advisory notice to insurers.

4. Use its building control powers to require properties in flood hazard areas to be constructed to resilient flood and windstorm standards as laid down by the ABI in conjunction with CML, RICS, and NHBC.

5. Consult with adjoining local authorities as necessary.

6. Adequately maintain watercourses and culverts using funding from a "planning gain" levy on developers, and publish biennial reports on actions taken.

7. Seek SUDS solutions to drainage of surface water, and agree to adopt and maintain above ground SUDS installations, again funded by a "planning gain" levy on developers.

Part three – National Government obligations

National Government to agree to:

1. Incorporate flood hazards and give statutory status to flood appraisal groups in legislation prepared under the Water Framework Directive.

2. Introduce regulations to enable local authorities to ring fence any planning gain levies to be used for maintenance of watercourses or SUDS schemes.

3. Introduce legislation to require resilient reinstatement following flood or storm damage in consultation with the insurance industry and CILA.

4. Temporarily relax FSA requirements on insurance companies' exposure management in local authority areas which have accepted the formula.

5. Make dams subject to COMAH Regulations.

6. Apply a levy on insurers to fund the introduction of the FASTER system for the collection of insurance company data on flood and storm claims and publication of analyses of these data. Require insurers to co-operate in the submission of such data, subject to confidentiality safeguards to protect personal information and insurers' competitive position.

8

New Strategies of Damage Reduction in Urban Areas Proned to Flood

Erik Pasche & Timm Ruben Geisler
Hamburg University of Technology (TUHH), Hamburg, Germany

ABSTRACT: In the past years an increase of flood hazards could be observed world-wide. While experts still discuss whether these hazards are the onset of a trend due to climate change or merely the consequence of cyclic changes there is a strong push on European politicians and decision-makers to reduce the vulnerability of urban areas along most of the big European rivers. Conventional flood management strategies which are targeting at the defence of these flood prone areas by dikes and walls are not an adequate answer. In general the investments for these defence structures can not be provided ad hoc. Additionally they will reduce the retention capacity of the river which needs to be compensated to avoid increased flood risk downstream. However possibilities for these compensation measures are limited. Looking for more economic and sustainable flood defence strategies, a great potential is seen in an improved flood resilience of urban areas. This comprises individual preventive and emergency measures at buildings and municipal infrastructure and a landuse policy to adapt building activities to the risk. It is the objective of this paper to illustrate this potential of flood resilience. Various techniques adapting buildings for flood will be shown, the preventible flood damage quantified and requirements as well as limits of flood resilience are described. Finally strategies of flood management are developed to make use of this potential.

8.1 RESILIENT FLOOD RISK MANAGEMENT

Several European countries have released a new water policy to cope with floods, e.g. (LAWA 1995, BMU 2004, IRMA 2002). Instead of fighting floods it gives preference to strategies of living with floods covering all measures of flood risk management by which the negative impact of flood is minimised. The underlying idea is to foster the ability of areas to recover after they have been exposed to flood which in a general sense represents the resilience of a system to persist and absorb changes and disturbances.

Land use control	Spatial planning
	   Flood risk adapted land use
	Building regulations
	   Building codes
	   Zoning ordinances
		Public responsibility

Risk awareness	Information:
	   Inundation maps
	   Risk maps
	Education:
	   Learning groups
	   Brochures
		Public and private responsibility

Flood preparedness	Flood resistant buildings
	   Waterproof materials
	   Sealing of buildings
	   Shielding of buildings
		Private responsibility

Financial preparedness	Obligatory insurance Private insurance of the remaining risk	Private responsibility

Figure 8.1. Sectors and measures of resilience in flood management.

A flood resilience policy aims at the minimisation of flood damages and the return of normal life as soon as possible after flood. In agreement to ecological systems in which the rate of resilience is determined by many factors, like variability of populations, their capability to adjust to the environmental conditions and to repopulate, flood resilience touches a variety of sectors and measures (Figure 8.1). They can be subdivided into four main groups of "land use control", "risk awareness", "flood preparedness" and "financial preparedness". The order of numbering corresponds to their relevance in flood resilience. Above all flood resilience is a matter of adequate land use. Thus spatial planning needs to consider flood aspects and has to define land use concepts for flood prone areas which includes building regulations, building codes and zone ordinances or even prohibits certain forms of land use.

Risk awareness comprises all measures which foster the risk perception of stake holders. While in the past water authorities and communities were afraid to inform stake holders about their flood risk, the new water policy aims at full and open information about flood risk. Inundation and risk maps generated by GIS have been found to improve the risk perception considerably (Figure 8.2) (literature). Other methods of risk perception are construction of flood marks/signposts, symbols and information centres.

Flood preparedness is mainly a matter of flood resilient building and hazard awareness. The flood resistance of buildings can be improved by holding back

```
┌─────────────────────────────────────────────────────────────────────────┐
│  LEVEL OF AWARENESS                                                        │
│                                                          Yes        No     │
│  Danger of flooding was known before the flood of 1993:  ☐         ☐      │
│  To the extend, that occurred in 1993:                   ☐         ☐      │
│                                                                            │
│                                                         1993       1995    │
│  Frequency of floods:            every 1 – 3 years       ☐         ☐      │
│  How often does a comparable     every 4 – 15 years      ☐         ☐      │
│  flood occur?            less often then every 15 years  ☐         ☐      │
│                                                                            │
│  Early warning time:             shorter than 1 day      ☐         ☐      │
│  How long prior to the occurrence did you  2 – 4 days    ☐         ☐      │
│  know the expected peek level and time?  Longer than 4 days ☐      ☐      │
│                                                                            │
│  Information:                       well informed        ☐         ☐      │
│  About the imminent flood      moderately informed       ☐         ☐      │
│  I / we have been ...               not informed         ☐         ☐      │
│                                                                            │
│  Experience with flooding:   Never experienced flooding prior to 1993 ☐   │
│                               Experienced flooding long time ago ☐         │
│                          Experienced last flooding shortly before ☐        │
│                                                                            │
│  MEASURES OF FLOOD PREPAREDNESS                                            │
│                                                                            │
│  Which measures of Flood Preparedness did you carry out before the floods in '93 and '95? │
│                                                                            │
│  Protection of movable fixtures                                           │
│  In the following columns pleas check in which manner movable fixtures have been protected │
│  against the water:                                                        │
│                                                         1993       1995    │
│  ...stored in the basement on raised elevation           ☐         ☐      │
│  ...brought from basement to upper storey                ☐         ☐      │
│  ...brought from basement to save location (outside the house) ☐   ☐      │
│                                                                            │
│  ...stored in the ground floor on raised elevation       ☐         ☐      │
│  ...brought from ground floor to upper storey            ☐         ☐      │
│  ...brought from ground floor to save location           ☐         ☐      │
│                                                                            │
│  Protection of immovable / built-in fixtures                              │
│  Please specify the articles and devices, which were rearranged or removed and in the │
│  columns below "measures" fill in the appropriate letters for the conducted measure for each │
│  year, e.g. "A" for "...dismounted (in basement) and stored on raised elevation". │
│                                                                            │
│  A    ...dismounted (in basement) and stored on raised elevation          │
│  B    ...dismounted (in basement) and brought to upper storey             │
│  C    ...dismounted (in basement) and brought to save location (outside the house) │
│                                                                            │
│  article / device                            measures 1993      1995      │
│  _____        ☐         ☐              │
│  _____        ☐         ☐              │
│  _____        ☐         ☐              │
└─────────────────────────────────────────────────────────────────────────┘
```

Figure 8.2. Resilient related questions from the questionnaire.

floodwaters through sealing the building or through barriers temporarily installed at some distance from a house (shielding). In case a flooding of the building can not be prevented the flood resilience of a building can be improved by using waterproof materials and evacuation of the water-sensitive interior and furnishing of the building. Sealing as well as shielding of buildings and evacuation of the interior requires sufficient preparation time. Thus measures to improve the flood resistance of buildings can only be effective if stakeholders are aware of the imminent hazard. This can be accomplished by flood forecasting and warning.

However their reliability is strongly influenced by the quality of monitoring systems and their promptness to transfer the recorded data to flood control centres.

Finally recovering after flood can be accelerated by damage compensation regulations and insurances.

The described measures of flood resilience need to be applied in an integrative way. Without land use control and risk awareness the efficiency of flood preparedness will be limited. Without risk awareness the readiness of stakeholders to invest in an improved flood resistance of their building is limited. Inconsistent land use can lead to inefficiency of flood preparedness or higher investments to accomplish flood resistance of the building.

In the last decade strategies and measures of flood resilience have been widely studied and published (IRMA-Sponge 2002, EA 2003). In England even a Kitemarking institution for approving techniques and materials of flood resilience has been established. In Germany many Governmental agencies and municipalities have delivered brochures to the stakeholders in which they inform about the risks of flood and possibilities to adapt their buildings to flood (MUF 1998, MURL 1999, BVBW 2003, EA 2003). However little is known about the efficiency of those measures either for individual properties or groups and total urban districts of communities. On this basis the benefit of flood resilience can not be evaluated in an economic way. The economic soundness of flood resilience compared to alternative flood defence concepts (e.g. conventional methods of defence by dikes and walls) can not be proved and stakeholders have difficulties to realise the payback and limits of their investments. The Hamburg University of Technology (TUHH), Germany has carried out an investigation in which the observed damages have been analysed in 9 communities along the river Rhine, Elbe, Danube and their tributaries with the objective to quantify the damage due to insufficient flood resilience. The work was done by order of the International Commission for the Protection of the River Rhine (ICPR) which published some of the results in (IKSR 2003). For the first time it could be quantified in what way flood awareness influences the construction of the fabric and interior of buildings and what damage can be reduced by different measures of flood resistance at buildings.

8.2   GENESIS OF FLOOD DAMAGE

The above mentioned study is based on a detailed analysis of the genesis of flood damages. With this knowledge about the reasons and ways of damage the weaknesses of existing buildings to resist water could be detected and the influence of risk awareness on the damage potential determined. Consequently the damage inventory of the study determined not only the total damage of the building but broke down the damage to the different floors and shows all affected items of the interior and at the fabric and specifies in what way they contributed to the overall damage. The inventory was mainly focussed on the direct damage. The indirect damages due to interruption of production, giving up of enterprises, reduction of property value or loss of unique cultural values has only been regarded in some few cases.

Most data resulted from inspection of the buildings and interviews with the users, making use of a questionnaire for which the resilient related questions are given in Figure 8.2.

All damage data were sorted into four main groups according to the location and source of damages which are damage at the *fabric* and *interior* of a building as well as the damage due to *cleaning* and *industrial/commercial* activity. While the fabric comprises the construction of the building (walls, ceilings, bottom-plate, window and door frames etc.) the interior includes damages at the fixtures, fittings and at the personal possessions.

From these data a resilience performance indicator C_R was derived which represents the damage ratio between the interior (D_I) and the fabric (D_F):

$$C_R = D_I/D_F$$

This parameter has been quantified for all analysed buildings and aggregated as mean value for the communities of Braubach, Kraiburg and Cologne (Figure 8.3). The dependency of this parameter on the level of flood resilience is clearly to be seen. While for buildings with strong risk perception the damage ratio between the interior and the fabric is small, for unprepared buildings the main source of

Attribute, location of damage	Experience	Region	Damage at the fabric D_F	Damage at the interior D_I	Resilience performance indicator C_R
Basement, new	Yes	Braubach	6.647 €	36.813 €	0,18
Basement, old	Yes	Braubach	8.692 €	34.257 €	0,25
Oil damage	Yes	Braubach	8.181 €	32.211 €	0,25
No oil damage	Yes	Braubach	10.737 €	33.745 €	0,32
-	Yes	Köln	9.203 €	39.369 €	0,23
-	No	Köln	17.895 €	30.166 €	0,59
Basement	No	Kraiburg	20.452 €	27.098 €	0,75
Oil damage	No	Kraiburg	23.008 €	27.098 €	0,85
Basement + ground floor	No	Kraiburg	25.565 €	24.031 €	1,06
-	No	Kraiburg	30.166 €	23.519 €	1,28
Ground floor	No	Kraiburg	33.745 €	12.782 €	2,64

Figure 8.3. The resilience indicator for different levels of risk awareness.

damage is the interior. Consequently low values occur for the indicator C_R at Braubach, a town at the Middle Rhine with regular flooding (every 2–5 years).

The indicator stays always below 0.32 whereas for the community of Kraiburg in Baden-Württemberg where a new settlement has been affected by flood the indicator is never below 0.75 and reaches values up to 2.64. Here the population had no experience of flood before the event of 1985.

The expert's survey in Braubach showed that in regions with high risk perception most buildings are without basements or inferior utilisation in the basement (empty or used as storage for properties which can be easily removed). Also in the first floor most fittings and furniture were temporarily removable. However rarely the fabric has been adjusted to flooding. In regions of little or no risk perception the utilisation of buildings is entirely different. According to Kraiburg nearly all houses have basements with superior utilisation (e.g. sauna, work bench, electrical appliances, boiler) and within the first floor most possessions and fittings have not been removed. Consequently this different utilisation of the building shows a strong feed back in the damage genesis.

Two different flood events of the City of Cologne give an idea on the change of flood resilience as a direct response to a experienced flood hazard. In 1993 the City of Cologne was hit by an extreme flood after a rather long period of no hazards. Thus many inhabitants were unprepared. Only two years later (1995) a further flood hazard occurred which was even more extreme. However this time the total flood damage was only 1/3 of the event of 1993. An analysis of the damage genesis showed that this difference in flood damage was due to a considerable decrease of damage at fittings and furniture. Consequently the flood resilience indicator changed from about 0.6 to less than 0.25. The low value corresponds to the one of Braubach. Obviously the population directly reacts on flood experience by adjusting the utilisation of the building. Adjustments at the fabric and fixations are only seldom to be found.

The invented damage data has also been analysed with the objective to identify the influence of oil contamination on the flood damage. In all surveyed communities oil contamination was observed but the impact was quite different. The source of oil contamination was always the oil heating. Either the flooded tanks were detached from their bracing by buoyancy effects braking off connecting pipes or water infiltrated into the tank through ventilation pipes and unsecured filling nozzles displacing the oil. The impact on the emitted oil on the environment varied in dependence on the available reaction time of oil. The faster the flood water moves the less is the effect of oil on the flood damage. This is proven by the observed flood damages at the communities of Garmisch-Patenkirchen and Baden-Baden where the mountainous rivers with strong current caused the flood. Here the observed oil emissions did not result in a distinct rise of flood damage. Otherwise in the community of Neustadt at the river Danube the oil emission was very harmful.

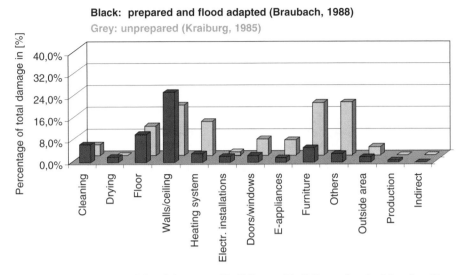

Figure 8.4. Comparison of flood damage of buildings with different level of flood resilience.

Figure 8.5. Photo of fabric damaged by oil.

Here a dike failed and flooded nearly the whole community. As the water could not flow off in this polder it stayed for several days with a water depth of up to 4,0 meters. Due to this extreme flooding many oil tanks collapsed. The polluted flood water penetrated into the fabric and walls, destroyed the thermal insulation and even materials which used to be resistant to water, e.g. tiles and stone floors. Most of the fabric and interior which had contact with the contaminated water could not be cleaned or repaired but had to be replaced (Figure 8.5). The damage of

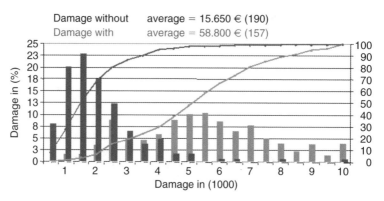

Figure 8.6. Flood damage affected by hazardous substances.

these buildings was more than three times of the ones without oil contamination (Figure 8.6).

This ratio is independent of the water depth and type of building. This result has already influenced the water policy in Bavaria/Germany. In the meantime all stake holders of buildings in natural inundation areas are obliged by law to pass a security check for their oil tanks done by authorised experts.

From this survey of flood damages it can be concluded that only the utilisation of houses is adjusted noticeably to an experienced flood hazard. Little effect can be seen at the fabric and fixations. This behaviour may be due to the investments needed to rise the flood resistance of the building. But discussions with stake holders demonstrated that they are in general not informed about the possibilities to improve the flood resistance of their buildings and about the economic benefits. And indeed we are still at the beginning of a technological development for the fabrication of water resistant buildings. The following chapter gives an overview of the state of the art in the technique of water proofing of buildings and shows deficiencies and limits which should stimulate researchers and engineers.

8.3   FLOOD RESISTANCE TECHNIQUES AT BUILDINGS

According to *Figure 1* the flood resistance of buildings can be accomplished by three defence strategies: *using waterproof materials*, *sealing* and *shielding the building*. In the first strategy the floodwater is not held back but the impact of water on fabric and fixtures is minimised. This strategy requires that all household products and furniture are removable. The second and third strategy try to keep the flood water out of the building. Hence they are often called dry-proofing strategy.

8.3.1   Water-proofing strategy

Walls and ceiling need to be sealed so that no water can penetrate into the structure. As in Germany the buildings are mainly build out of stone and concrete the survey of appropriate flood resilient techniques has been mainly concentrated on solid buildings. Thus the presented techniques are restricted to that type of building.

For a solid masonry a water-resistant fabric can be easily accomplished as long as it is a new building. Then water-tight concrete can be used for walls and ceilings. Water-tight brick stones are also possible but at joints there is always a risk of crack formation. At existing solid buildings walls and ceilings are often composed of porous material and thus need a coating to become water-resistant. At the inside face of an external wall an appropriate mean would be a polymer bituminous seal (damp proof membrane) or a render of sand/cement between wall and plaster. Gypsum based plaster or plasterboards which are filled with gypsum have to be avoided. Plasters of synthetic resin, hydraulic lime coatings and special plasters on the basis of plastic show good water resistance. Ceramic tiles can also provide a water-resistant surface.

At the outside face of the external walls cracks should be repaired and the brickwork coated by a water resistant paint or render. The external insulation can be adapted to flooding by using self-draining mineral fibre batts or boards or rigid plastic insulation. These materials will dry out after flood without damage. All coatings at the inner and outside face of the external walls should be applied to 500 mm above the maximum expected level of flooding or the capillary uprise is stopped by a capillary barrier out of epoxidharz (damp proof course). Any measures to improve water resistance at walls must allow adequate water vapour transmission to avoid trapping moisture within the wall. The seal should not be penetrated by cables, pipes and fixing bolts. External water supply pipes, drain pipes, electricity supply and telephone cables need to be conducted through the masonry and must be secured by sealing sleeves on both sides of the masonry and conduits of plastic or stainless steal.

For the floor similar water-proofing measures are necessary at the concrete slab. Cracks need to be repaired and a polymer bituminous seal should be arranged as damp proof membrane between the slab and the screed. A seal between the soil and slab is only feasible for new buildings but it is not necessary if the concrete is tight enough. At the joint between the masonry wall and the floor the seals must overlap to acquire effective connections. Additionally a damp proof course has to be installed in the walls and to be connected with the outside and inside seal of the external walls to prevent water seeping up from the ground.

Wood, paper, textile and carpets are inconsistent with water. Thus the plaster should not be coated with wall papers. No chipboards, parquets and fitted carpets should be found on the ground floor. An underfloor heating system may be damaged by flooding and should not be applied. Thus the floor should have a finish of tiles out of ceramic, terracotta or other non-porous material. The thermic insulation below the finish should consist of rigid boards with low water absorption. For acoustic insulation the concrete slabs of the floor are covered with an additional layer of concrete screed which should also be water-resistant. Despite the use of water-resistant materials seepage through the screed can not be totally avoided. Thus it is recommended to arrange a gap by a drainage layer between the polymer bituminous seal and the screed so that air can circulate freely speeding up the drying process. In basements and ground floors sumps for electrical pumps should be installed.

Fittings should be removable or withstand water without damage. Toilets and hand basins are not normally affected by flooding. Doors, windows, skirting boards and staircases are only slightly affected if they are made of solid timber, PVC and aluminium. The electricity supply system can be adjusted to flooding by locating the electricity meter, consumer unit (fuse box), the ring main cables and the sockets to a level above the expected flood level. Normally modern electrical wiring is not affected by immersion in floodwater. But it is better to use plastic cable conduits in order to reduce the cost of re-wiring. Gas and oil fired boilers, burners and associated pumps are very sensitive to floodwater and need to be located above the expected flood level. Oil tanks need to be fixed at the masonry. However the masonry has to be able to take the forces resulting of buoyancy of an empty tank. Otherwise the tank has to be relocated at a higher level or the oil fired heating system has to be replaced by gas driven system. Additionally ventilation pipes and filling nozzles need to be equipped with water-tight caps. Finally drains and sewers need to be closed temporarily during flood by non-return valves (flap gates) because the backflow of sewage can cause blockages and additional pollution.

8.3.2   Dry-proofing strategy

In the dry-proofing strategy measures are taken to keep the water out of the house. They include the installation of temporary flood barriers, measures to reduce seepage through walls and floors and the installation of non-return valves on sewers and drains. Two different techniques are applied for the dry-proofing of buildings, the *sealing* and *shielding*.

In the first case the external wall of the building and the ground floor are used to hold back the flood water. Thus they must be sealed at the outside face similar to the water-proofing of the masonry. Only for new buildings the ground floor slab can be sealed from the outside. The refurnishment of old buildings with a water-tight ground floor requires either the replacement of the existing concrete slab or its covering by a reinforced and water-tight concrete cap (Figure 8.7). The weight of the floodwater (water pressure) might require an enforcement of the external walls. In general this is necessary for a flood stage which is more than 1 metre

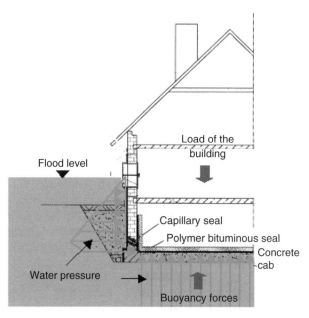

Figure 8.7. Dry-proofing of a building by sealing.

above the deepest part of the dry-proof building. In these cases the concrete cap need to be extended over the inside face of the external walls with a height corresponding to the static requirements.

Below this concrete cap a damp proof membrane out of polymer bitumen further reduces the risk of seepage. The damp proof course in the external wall needs to be connected with the outside seal of the external walls and the damp proof membrane (Figure 8.7) to prevent the ingress of water through the external wall from the ground. The construction of this concrete cab is cost-intensive and thus not always feasible. In these cases a partial flooding could be a good alternative to water-proofing and total flooding (Figure 8.8). Through a valve controlled pipe in the external wall deep sections of the building are filled with flood water. The valve, a pump in the sump and pressure sensors will regulate and control the water stage within the building that the water pressure is kept below the critical load of the external walls and ground floor This technique is recommended especially for buildings with cellars. Water-proofing at the cellar and dry-proofing at the first floor are applied in combination and ensure that floodwater can be kept out of the first floor as long as the floodwater level stays below 1 metre of the first floor.

Doorways, windows and air bricks need to be closed with temporary flood barriers. For doorways the barrier is often composed of hollow aluminium beams which are stacked on top of each other. At the upside of each beam a seal is attached. The bottom beam has a special seal at the downside. The beams are

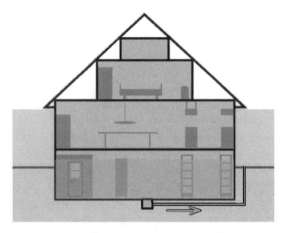

Figure 8.8. Partial flooding of dry-proofing buildings.

Figure 8.9. Temporary barrier of a doorway.

installed in a frame which is permanently fixed at the masonry. They need to be pressed together and horizontally stabilised in the frame. This can be accomplished with a sledge which can ride vertically in the frame and can be screwed with the frame and beams (Figure 8.9). At the windows the same system is applied. But often plastic or metal boards are applied instead of beams.

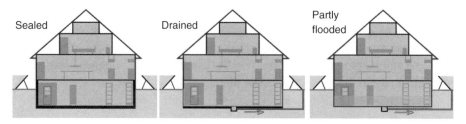

Figure 8.10. Dry-proofing of houses by shielding.

Figure 8.11. Free-standing inclined flood barrier.

In case of *shielding* the removable flood barriers are installed at some distance from the building or a group of properties. The floodwater does not reach the building itself. However in dependence on the permeability of the underground lower foundations, ground floor levels and cellars need to be sealed, drained or partly flooded (Figure 8.10). The removable barrier solution described above can be used for the shielding as well. But instead of a frame the beams are ending at pillars which are temporarily fixed at a concrete bearing plate which surrounds the building. Cheaper solutions are free-standing barriers which need no bearing plate. They can take the floodwater load without anchorage by creating pressure and friction between the system and the ground. However the ground must be firm and accept seepage without hydraulic erosion (Figure 8.11). Several systems are available ranging from rubber belts, fiberplastic coats to inclined barriers out of standard palates and foldable metal support. The height of the free-standing barriers should not exceed 1 metre although even higher systems are available. The risk of failure due to underground instability is high. Subsoil erosion and hydraulic blow have to be studied prior to the application of free-standing barriers. Most sensitive is the shielding technique to waterways in the underground. Backflow through sewers and drains have to be stopped by non-return valves. Supply lines under free-standing barriers need adjustment to hinder seepage flow along cables and pipes.

8.4   EFFICIENCY OF RESILIENCE MEASURES

Only little experience exists about the application and efficiency of flood resilience measures at buildings. A systematic application to a whole region or street has not been done yet. This need to be co-ordinated by central organisations like water boards. But they leave the application of these measures totally to the individual initiative of the stake holders.

The flood damage inventory of this study carried out at 9 different locations in Germany showed only few buildings with structural resilience. On this basis it is not possible to draw general conclusions on the costs and benefits of flood resilience. Thus the efficiency of flood resilience was derived in an indirect way. At all invented buildings the different resilience strategies were applied theoretically, the avoided damage evaluated and normalised over the total damage. Averaging in dependence on the resilience strategy and risk awareness resulted in the efficiency numbers of (Figure 8.12). Clearly to be seen is the strong dependency of risk perception on the efficiency of water-proofing measures. While for regions with good flood risk awareness the total damage can be reduced by 7–20% through adapted use and removal of furnishings regions of no flood risk awareness reaches efficiency numbers of up to 60%. In contrary the efficiency of dry-proofing measures show only very little influence on the level of risk awareness. This underlines the little readiness of the stake holders to take dry-proofing measures. This result surprises with regard to the high efficiency of dry-proofing measures (up to 83%). One hundred percent efficiency is theoretically feasible for dry-proofing strategies. As the height of free-standing barriers and sealing of the external walls of the building is limited to about 1 metre several of the invented buildings could not be protected by this resilience strategy. Consequently the average efficiency of the dry-proofing strategy was reduced for the total of invented buildings.

Incomplete is the knowledge about the efficiency of resilience measures at buildings with industrial business. Only data from small business objects have been available in a sufficient number that general conclusions could be drawn (Figure 8.13). These buildings behave similar to domestic houses but the efficiency can vary strongly in dependence on the type of industrial activity. In regions of no risk awareness the flood damage of small business objects is dominated by damage at appliances and furniture which can be up to 70%. By water-proofing and improved mobility this damage potential can be totally avoided. However in regions of good

Resilience strategy	Degree of risk awareness		
	No awareness		Good awareness
	Without warning	With warning	With warning
Water-proofing			
Water-proof material	15–30%		10–25%
Flood adapted use	30–40%		7–10%
Removal of furnishings	50–60%	30%	15–20%
Dry-proofing			
Shielding of building	26–75%		23–70%
Sealing of building	40–83%		30–75%

Figure 8.12. Efficiency of flood resilience measures at domestic homes.

Resilience strategy	Degree of risk awareness		
	No awareness		Good awareness
	Without warning	With warning	Without warning
Water proofing			
Water proofed material	10–20%		10–35%
Flood adapted use	10–15%		13–18%
Removal of furnishings and industrial appliances	20–70%	?	5–25%
Dry-proofing			
Shielding of building	25–82%		10–77%
Sealing of building	55–90%		40–80%

Figure 8.13. Efficiency of resilience measures at buildings with small business.

flood awareness this damage source is already down to 5–25% that for theses buildings dry-proofing measures become more efficient.

8.5   CONCLUSION AND OUTLOOK

The different techniques for the improvement of flood resilience at buildings has been analysed and their efficiency in reducing the flood damage has been evaluated.

A resilience performance indicator C_R has been introduced which represents the damage ratio between the interior (D_I) and the fabric (D_F). This indicator was found to be a good instrument to quantify the risk awareness of the population and the level of flood awareness. For good flood awareness this parameter has values which are less than 0,30 whereas for regions with little flood awareness the resilience performance indicator reaches values of up to 1,0 and higher.

Stake holders react on experienced flood by improving the mobility of their furniture and appliances. Measures at the fabric to withstand floodwater are only rarely taken. Thus the great potential of flood resilience is to be seen in the damage reduction through water-proofing and dry-proofing. The necessary measures and available techniques for these resilience strategies have been explained with special reference to buildings with solid masonry, the dominating construction type of houses in Germany. Dry-proofing can be accomplished by sealing or shielding the buildings. The limits of this technique are marked by the stability of the building. In general flood stages of more than 1 metre above the ground floor level can not be taken by the masonry. Enforcements by capping the ground floor and the external walls at the inner side are appropriate measures to overcome these stability problems. However they are cost-intensive. An alternative can be the partial flooding of the building. Especially the flooding of cellars is an economic way to generate the necessary counterproof to the floodwater. However the partial flooding has to be controlled and additionally to the dry-proofing measures the flooded part of the building requires wet-proofing.

The efficiency of these resilience measures have been determined by applying them to the invented buildings in 9 German flood regions. Theoretically a "one-hundred-percent-efficiency" can be accomplished. However for many buildings the flood stage was 1 metre and higher that they need to be partially flooded. Thus the average efficiency for all analysed buildings was about 80% to 90%.

Despite this great resilience potential of dry-proofing and water-proofing the readiness of the stake holders has been low to apply these techniques. The reasons are to be seen in insufficient information about the techniques and the economic implications. It is difficult for the stake holders to assess the economic efficiency of these resilience measures.

To overcome this reservation of the stake holders more studies are necessary in which the costs for dry-proofing and water-proofing are determined for different boundary conditions of the buildings (age of the building, material of the masonry, flood stages, etc.). The study should be extended to other European countries to cover the full scope of construction types of buildings. It is further recommended that the water policy should create the administrative and legislative framework that flood resilience measures are co-ordinated by water authorities and the stake holders are financially supported. An internet-based resilience information system (e-RIS) could be a good instrument to support the water boards in their policy. It should inform the stake holders about their flood risk (risk perception enhancement), provide an interface by which the affected population can enter their building and living condition into the system and an expert shell which informs about appropriate resilience measures in dependence of the input data. Simultaneously the water boards are informed about the damage potential in their flood regions, can analyse different flood defence strategies (technically) and derive the appropriate flood management strategy by a cost-benefit analysis.

REFERENCES

BMU. Berlin 2002. 5-Punkte-Programm der Bundesregierung: *Arbeitsschritte zur Verbesserung des vorbeugenden Hochwasserschutzes, Bundesministerium für Umwelt, Naturschutz und Reaktorsicherheit.* http://www.bmvbw.de/Anlage12654/5-Punkte-Programm-der-Bundesregierung.pdf

BVBW. Februar 2003. Hochwasserschutzfibel, *Planen und Bauen von Gebäuden in hochwassergefährde ten Gebieten,* Bundesministerium für Verkehr, Bau- und Wohnungswesen, Berlin. http://www.bmvbw.de/Anlage16214/ Hochwasserschutzfibel.pdf

EA. May 2003. Flood Products, *Using Flood Protection Products – A guide for homeowners,* Environment Agency. http://www.environment-agency.gov.uk/commondata/105385/flood_product_guide_lowres.pdf

IKSR. Koblenz 2002. Hochwasservorsorge – *Maßnahmen und ihre Wirksamkeit,* Internationale Kommissi on zum Schutz des Rheines (IKSR). http://www2.ms-visucom.de/r30/vc_content/bilder/firma20/pdf/RZ_iksr_dt.pdf

IRMA. Delft 2002. *Towards Sustainable Flood Risk Management in the Rhine and Meuse River Basins,* IRMA-Sponge, Netherlands Centre for River Studies. http://www.irma-sponge.org/

LAWA. Stuttgart 1995. *Leitlinie für einen zukunftsweisenden Hochwasserschutz, Länderarbeitsgemeinschaft Wasser, Umweltministerium Baden-Württemberg.* http://www.lawa.de/pub/kostenlos/ww/Messdienst.pdf

MUF. Mainz 1998. Hochwasserhandbuch, *Leben, Wohnen und Bauen in hochwassergefährdeten Gebieten,* Ministerium für Umwelt und Forsten Rheinland-Pfalz. http://www.wasser.rlp.de/download/Hochwasserhandbuch.pdf

MURL. Düsseldorf 1999. Hochwasserfibel, *Bauvorsorge in Hochwassergefährdeten Gebieten,* Ministerium für Umwelt, Raumordnung und Landwirtschaft, NRW. http://www.lua.nrw.de/wasser/hochwasserfibel.pdf

9

Flood Resilience in the Built Environment: Damage and Repair

Stephen Garvin
Building Research Establishment, East Kilbride, Scotland, UK

ABSTRACT: Flooding can cause serious damage to buildings, resulting in the need for extensive repairs to be undertaken. The amount of damage depends on various factors such as the cause, depth and duration of the flood. Repairs to flooded buildings should seek to increase the flood resilience and thereby ensure that in any future flood the amount of damage is minimised.

This paper sets out the type of damage caused to different building elements, the assessment of future flood risk and the repair options. Repairs are in accordance with the level of risk involved and the costs involved. Improving the flood resilience of buildings should result in reduced insurance costs.

9.1 INTRODUCTION

Recent flooding events across Europe have shown the devastating impact that flooding can have on people, property and business (ABI, 2003; The Environment Agency, 2001). Although, it is unlikely that the effects of flooding can be eliminated completely, many practical steps can be taken to reduce the cost of flood damage to buildings by improving flood resilience. This paper will look at the impact of flood on buildings and the approach to the repair of existing buildings.

Flooding can occur as a result of the following (ODPM, 2002; CIRIA, 2003):
- flooding from watercourses such as rivers and streams, associated with extreme rainfall, snowmelt or hail;
- surface water run-off from hills;
- the sea, involving overtopping of sea defences;
- groundwater rising into buildings;
- infrastructure failure;
- blocked or overloaded storm-water drainage systems and sewer flooding.

The sources and causes of floods affect the damage to the building and the means of reinstatement and repair (BRE Report BR466).

The insurance of buildings against flood damage varies across Europe, with some countries offering a mix of public and private insurance. In the UK flood insurance is entirely private and the emphasis is to ensure that new buildings are not built in flood prone areas. This does not deal with all the potential sources of flooding and the reality that flood risk already exists to many buildings. Increasingly the flood resilience of buildings in flood should be seen as a means to secure further insurance at a reasonable cost of at risk buildings.

Preventing flood water from entering buildings should be the first consideration. However, there is unlikely to be absolute certainty that floods can be avoided. Improving the flood resilience of buildings is likely to include either the wet-proofing or dry-proofing of the existing buildings.

9.2   IMPACT OF FLOODING ON BUILDINGS

The National Trust for Historic Buildings (USA) describes the impact of water on buildings as follows:
– "No other 'element' is as destructive to buildings as water."

The effects can be described as follows:
– direct deterioration of building materials susceptible to water.
– longer term deterioration or effects due to higher levels of moisture than acceptable for durability.
– secondary deterioration due to freeze/thaw damage or salt crystallisation.
– physical damage caused by weight of water or flowing water.

Direct effects of water will be immediately obvious, but other longer term and secondary effects will also occur unless measures are taken to prevent future risk. Proper drying out of buildings is essential to prevent any risks.

The insurance costs to repair damage to a building are in the region of £15,000 to £30,000 as a result of direct damage to the building. In addition, £9,000 could be added for damage to building contents. It is therefore preferable to reduce the potential for damage by using materials that have inherent durability to moisture.

The amount of damage that is likely to be caused to buildings will depend in part upon the water depth that is reached and the duration of exposure. All buildings, which have not been dry-proofed, will allow water to enter during a flood through the following routes:
– permeable masonry, including through mortar joints;
– vents and air bricks;
– windows and doors;
– service outlets;
– pipework;
– ventilated openings in walls;
– basements and under ground floors.

The following indicates the likely degree of damage under different depths of flood:

- below ground floor – minimal damage to the main building, damage to electrical sockets and other services in basements, carpets and other fittings in basements, possessions in basements. Deterioration of floors may result if the flood is of long duration, and drying out is not effectively undertaken. Buildings with level access thresholds may allow shallow depth flooding to enter at the front door to the floor and finishes.

- up to one metre above ground floor – damage to internal finishes, saturated floors and walls, damp problems result, chipboard flooring destroyed, plaster and plasterboard likely to deteriorate. Services, carpets, white goods, furniture, electrical goods and belongings all likely to be damaged to point of destruction.

- more than one metre above ground floor level – damage to walls and structural damage. Services such as water tanks and electrical services may be damaged.

9.3   DAMAGE FROM FLOODS

In repairing buildings in flood prone areas it is important to consider the design and materials used. The building elements described in this section need to be given careful consideration.

9.3.1   External walls

External walls are generally built from solid masonry, cavity wall masonry, a frame construction with infill masonry or covering cladding (BRE Report 352). The wall will include various materials such as masonry units (clay brick, concrete block, stone, etc), mortars (cement or lime based), ancillary components (wall ties, etc), external renders, internal plasters, plasterboard, cladding materials (steel, aluminium, clay, etc), fixings and sealant. Damp proof courses and insulation will also be present in many modern walls, but they may not be present in older properties.

The exposed components of the external wall need to be resistant to the effects of water in any building in order to prevent damage from rainfall and driving rain. However, bricks and blocks can be saturated by flood water and may suffer from secondary effects such as freeze/thaw damage if freezing weather follows the flood. Efflorescence and lime bloom may result from flooding as the materials dry, this is not a deterioration mechanism, but can be difficult to remove. Cement or lime based renders may be susceptible to detachment, especially if exposed to frost or rapid drying out soon after the flood. Cladding and timber finishes could be subject to loss of coatings and risk deterioration or corrosion.

The other components of the wall, such as wall ties and internal plasters that are not exposed to water in normal use could be susceptible to damage in a flood.

Gypsum plaster may be damage in relatively short flood durations, but lime plasters should be more resilient. Metallic components should not be left wet for any period of time as corrosion may start. In addition, timber wall components can begin to decay if their moisture content remains above 20% for any period of time.

The different types of insulation materials used in either cavities, internal or external insulation varies in its resilience to flooding. Loose fill insulation (that is generally blown into the cavity) and wools will typically be effected by water, whilst closed cells plastics are unlikely to be affected. The loose fills and wool types may slump in wall cavities should they become wet, rendering them useless. Smaller amounts of water could affect the insulation properties.

The flood water can affect the structural integrity and durability of the external walls. Walls that are saturated for long periods of time could be left susceptible to frost damage. Water that breaches the damp proof course level can rise into the masonry further than the level of the flood itself.

9.3.2   Internal walls

Internal walls can be built from masonry or framed structures (including timber stud or lightweight metal frame). On masonry they are generally covered with hard plaster that is applied direct to the masonry, or plasterboard that is fixed on plaster dabs or onto timber supports. Framed walls are generally covered by plasterboard that is fixed directly to the frame. Plasterboard, gypsum plaster and decorative finishes can be destroyed in flood. If the materials have aged or been subject to heating then they will rapidly deteriorate in a flood.

Lime based plasters are more resistant to flood. Impermeable wall finishes such as tiles are the most resistant to flood water and prevent the internal leaf of cavity walls from becoming saturated. Plasterboard can be placed horizontally onto the wall in the event of a flood the plasterboard at lower level may need to be replaced, but not at higher level.

Insulation in the wall cavity can be destroyed in the same way as for external walls. Similarly timber and metallic frames are at risk if they remain wet for extended periods of time.

9.3.3   Floors

Floors are generally either timber suspended, suspended concrete or solid concrete. In flood prone areas the solid concrete floors are generally considered more resistant (BRE Report 332; BRE Report BR440).

They prevent water being allowed to fill an under floor void in the same way that the suspended floors allow. However, the concrete has to be of good quality and the proper installation of the damp proof membrane is essential, with connection to the wall damp proof course. The main risk is damage to the floor screed in the event of a flood, this would result in the screed needing to be lifted and re-laid.

Insulation under the floor will be at risk if the water enters around it through cracks or joints, or damaged screed finish. Wool type insulation may be damaged, although closed cell types should not be affected by short term water contact. Perhaps the main issues is the problem of drying the insulation and ensuring its thermal performance is not affected.

Timber floors including the floor boards, joists and other timber components will be at risk if the moisture content is raised above 20% for a period of time. Some types of floor finishes, such as chipboard, can be damaged by the short term water contact and may be rendered unusable. Susceptible floor covers should be avoided where flooding of buildings is expected. Effective drying of timber floors and proper monitoring of moisture contents is essential if durability problems are not to occur. It should, however, be appreciated that the softwoods used to construct the floor joists will not be damaged immediately by water contact and should not suffer long term damage if dried properly.

9.3.4   Services

Electrical wiring, gas supplies, pipework, heaters and other services should be protected as far as possible from water. The actual impact of the water will vary depending on the service.

9.3.5   Fenestration

Fenestration covers external doors, windows and glass (and associated hardware). These items are not generally affected by short-term exposure to flood water. However, leakage of water can occur through seals and vents, this allows water to access the inside of the building and lead to damage to internal fixtures, fittings and materials.

Glass breakage can occur if water pressure increases substantially on the window. Single glazed windows, with annealed glass, would be particularly at risk, whilst double-glazing units with toughened glass should be more resistant to breakage.

Coatings on timber frames and doors may deteriorate if long term exposure is experienced. Timber may absorb water leading to distortion and cracking as they swell and during subsequent drying out. External joinery needs to be protected, in the main, from water contact with good coatings and good joinery work.

9.3.6   Flood resilience of buildings

It is clear from the proceeding discussion that building elements can be susceptible to the effect of flooding. In designing new buildings or adapting existing buildings to resist the effects of flooding it is necessary to both design to resist water and use materials that have natural resistance, or dry quickly with minimal intervention after a flood.

The construction of buildings that are fully flood resistant, i.e. have no potential to be damage or leak water to the inside would be prohibitively expensive. It is preferable to consider how to make the buildings resilient, so that if a flood occurs then the impacts are minimised and any damage caused can be readily repaired.

For existing buildings there is potentially less adaptation possible. Walls, floors and fenestration may be fixed. However, if a building is flooded and requires repair it is preferable to increase the buildings natural resilience. However, this should be done on the basis of a risk assessment that will take account of the future flood risk. The level of risk should dictate the nature of the repair and whether or not it is necessary to improve the flood resilience. There are in most cases cost implications of improving the flood resilience, however, where there is a high risk of a flood returning then this should be paid back.

9.4   DECONTAMINATION AND DRYING

Decontamination should begin as soon as possible after the flood recedes, preferably within 24 hours. The decontamination of a flooded building can be fully effective, especially where the contamination is restricted to accessible areas. However, there are a number of circumstances in which it will be necessary to undertake further action. Wall and floor cavities can be particularly difficult to clean, especially where they contain insulation.

Drying should follow on immediately from the decontamination process. It is preferable if drying can be achieved without opening up of walls or floors, however, this may be unavoidable. In this case reinstatement of finishes should be delayed until after the moisture content of the material falls sufficiently, otherwise damage and mould can result.

Specialist contractors should be employed for the decontamination and drying of the buildings after flooding.

9.5   FLOOD RISK ASSESSMENT AND POST FLOOD SURVEY

Undertake flood risk assessments to determine the repair requirements of flooded buildings.

9.5.1   Risk assessment procedure

A risk assessment procedure will involve identification of the potential flooding hazards, followed by assessing the risk.

9.5.2   Hazard assessment

Table 9.1 provides a list of risk factors that should be assessed. In addition, add any additional information that is not covered in this list for specific situations.

Table 9.1. Hazard assessment issues for flood risk.

Risk Factors

If a flood zone map is available is the building within the Indicative High Risk flood plain zone

If a flood zone map is available is the building within the Indicative Medium Risk flood plain zone

If a flood zone map is not available is the building situated on alluvium

If a flood map is not available, is the building below 10 m AOD

Have flood warnings been issued for the area

Is the Local Authority aware of any historical flooding of this building

Do the characteristics of the site suggest that it is prone to flooding

Is the building underlain by a chalk aquifer

Is the building close to a winterbourne stream

Is the building within a natural or artificial hollow, at the base of a valley or bottom of a hillside

Does the building impede natural or artificial land drainage flow paths

Are there signs of soil erosion upslope of the site

Does the land upslope of the site suggest that overland flow will be encouraged

Are there artificial drainage systems on, adjacent to or upslope of the site

Is the building protected by a flood or coastal defence

Is the development protected by a flood control structure (e.g. flap valve, sluice gate, tidal barrier, etc)

Is the building located upstream of a culvert which may be prone to blockage

Are water levels in a watercourse located next to the building controlled by a pumping station

Is the building adjacent to or downstream/downslope of a canal

Is the building downstream/downslope of a reservoir or other significant water body

Has sewer flooding occurred as a result of blocked and poorly managed sewers

Is local rainfall in excess of 750 mm per year; is climate change likely to increase the level

Is there a high ratio of hard standing to soft ground (greater than 1:1 ratio)

Is the soil impermeable or often near saturation

Is the infrastructure well maintained for drainage, sewers and water capacity

Does local knowledge indicate the building is likely to flood, e.g. past history or road/house names that suggest flood history*

Determine what occurred when the flood took place through the site survey, including the following:

– the depth of the flood relative to the building and any differences between the external depth and internal depth of the flood;

– the duration of the flood;

– the type of flood that has occurred;

– the elements of the building affected;

– the degree of damage caused to the building elements, and differentiation made between material and structural damage;

- the probability of future floods occurring, taking into account the impact of climate change, which is likely to increase the risk (FBE Report No. 2, 2001);
- the extent and standard of existing flood protections and their anticipated effectiveness over time;
- the rates of flow of the flood water, which will affect the potential for structural damage and the ability to erect temporary flood protections;
- the remaining lifetime of the building and the extent to which it is currently designed and built to deal with flood risk.

9.5.3   Assessing and managing the risk of future flooding

An assessment of the risk of future flooding is essential to the repair of the flooded building. In some cases the building may have been extensively damaged, however, the risk may be assessed as sufficiently low of a future flood that no increase in the resilience of the building needs to be considered.

Three categories of flood risk apply, which are described as follows (for each flood risk there is a corresponding standard of repair):

- little or no risk, i.e. the lowest level, effectively where there are no risk factors, or those determined are being managed effectively. The repair of the building can be in line with the original specification. The specifier should be aware that the flood resilience will therefore not be improved and that this option is likely to be used in limited circumstances of buildings that have been flooded. The specifier will need to be certain that the building is not at risk of future flooding.

- low to medium risk, risks will have been identified through the hazard assessment. Sufficient risk factors will have been identified that an increase in flood resilience will be required. Any damaged building elements will be repaired to a higher standard of flood resilience.

- high risk, significant risks will have been determined from the hazard assessment. The potential for a repeat flood will be considered high and the consequences of such a flood severe. Structural damage will have resulted from the flood and this will require an increase in the resilience of the building. Structural strengthening will be required to ensure that repairs are successful.

Inevitably in assessing the risk of future flooding there will be a degree of uncertainty involved. Where there is significant uncertainty then assessors are advised to take exercise caution and to assign a higher level of risk.

9.6   REPAIR FOR RESILIENCE

As the flood risk increases to a building then so should the specification of the repair of the building to improve the resilience. Repairs should concentrate on the affected buildings elements, although for some buildings it may be preferable to undertake repairs to unaffected elements if there is a high risk of a future flood. This should be done whilst affected elements are being repaired (BRE GRG 11).

Improving the flood resilience of buildings can include both dry-proofing or wet-proofing measures, which are defined as follows:

– dry-proofing involves the use of barriers to prevent water entering doors, windows, air-bricks, ventilation holes and other openings (CIRIA, 2003). Dry-proofing is intended to prevent water from entering the building and thus causing wetting and damage to building materials. Dry proofing involves the use of flood protection products that can be bought as proprietary products from manufacturers. Building owners can also use sand bags or plastic sheeting to protect doors or walls from flood water.

– wet-proofing measures include the use of flood resilient building materials for walls and floors and in other parts of the structure, and the raising of electrical wiring above flood levels. In wet-proofing there is an acceptance that some water will enter the building and the intention is that the design and/or materials used help to prevent any damage to the building.

Often an appropriate combination of dry-proofing and wet-proofing measures offers the most effective solution, especially where there is a high risk of flooding. However, where flood waters rise above one metre depth, dry-proofing measures become undesirable as the pressure of holding the floodwater back can cause structural damage to the building, dry proofing measures should generally be limited to below this level.

The repairs specified for the flooded building will be determined from the post flood survey and the flood risk assessment. As the risk of flooding increases then the specification of the repair will be increased. Repairs can therefore be categorised into three levels that correspond to the risk levels. i.e. Little or No, Low to Medium or High.

9.6.1   Repair – little or no risk

The risk assessment will have shown that there is little or no risk of flooding. The repairs to the building will involve little more than the original specification. However, measures are suggested to reduce the potential for water leakage through external walls, such as improving the sealing around windows and doors, or the application of a water repellent to the brickwork up to a height of one metre. Internally measures can be taken to improve the sealing of the floor to prevent water ingress, for example, by applying sealant to the junction of walls and floors.

9.6.2   Repair – low to medium risk

The risk assessment shows that the likelihood of the flood returning is such, i.e. low to medium, that increasing the resilience of the building is essential. The repairs to the building will involve increasing the resilience above the original specification.

This will involve measures to ensure that the materials used in the building are not at risk of damage from water, and that water leakage into the building is reduced. Examples of such measures would involve repairs to the external brickwork of walls. Here damage may not be caused directly to the bricks, but water may leak through poor quality masonry particularly through the mortar joints or the interface between the mortar and the brick. In addition, holes in the masonry, such as air vents, weep holes and service entries will leak water to the cavity or inside of the building. Repair options for the walls might include re-pointing the mortar to reduce leakage or applying an external render (BS5262, BRE Digest 410, BRE GBG 18). The latter option has distinct impact on the appearance of the building as well as the potential to improve its performance. There may therefore be aesthetic issues to be determined with the local authority or others.

For internal blockwork walls where plasterboard is used the options would include replacing the plasterboard with wet applied plaster directly to the wall, or using plasterboard hung horizontally. A good quality lime : cement plaster could be used and would be resistant to the effects of floods (BS5492). However, it may not be possible to accommodate electrical services with the wet applied plaster. Therefore, the horizontally hung plasterboard may be a better option in such an instance. In the event of a flood the lower plasterboard run is lost, but not the whole height of the rooms.

Measures can be taken to avoid damage to services during flooding and these may be more realistic than seeking resilient solution. Examples of ways to manage the risks are as follows:

– place electrical sockets and wiring higher up the walls of ground floors;
– locate other services such as boilers should be on upper floors where possible;
– fit drainage services with valves that close to prevent back flow of flood water (NB this prevents use of the services until the flood retreats).

9.6.3 Repair – high risk

The risk assessment will have shown that the risk of flood is high and that increasing the resilience of the property is essential. Resilient measures include the following:

– in the case of structural damage to the building it will be necessary to undertake strengthening work on the building to ensure that any future flood does not have damaging effects. The structural damage may have resulted from the poor condition of building elements such as masonry walls or structural frames, therefore the repair may have been required in any case (BS5628: Part 3). Structural damage might not indicate a fault with design, materials or workmanship, but this should always be considered as a possibility.

– additional external protection methods could also be used to ensure that water is kept out of the building, i.e. its dry-proofing is improved. Items that could be

Table 9.2. Particular flood damage weaknesses of typical construction elements.

Construction	Particular weakness	Preferred solutions
Masonry walls	Damage to plasterboard Damage to gypsum plasters Damage to traditional lath and plaster finishes	Replace with wet applied render/ remedial waterproof plaster. Replace plasterboard with vertically hung boards to reduce the amount of damage in future floods.
Timber frame walls	Water damage to frame structure (including rot and warping)	Expose timbers subjected to flood waters to facilitate drying. Assess and replace timber where not fit for purpose. Where appropriate treat timbers. Ensure all timber is thoroughly dried before replacing finishes.
	Damage to plasterboard	Where only the lower part of the wall is affected, install plasterboard horizontally to avoid whole room replacement.
Steel or concrete frame walls	Corrosion risk to steel frame, or reinforcement	Ensure that the frame is exposed, cleaned and allowed to dry quickly after the flood.
	Damage to plasterboard	As for timber frame.
Timber suspended floors	Water damage to floor cover and structure (including rot and warping)	Remove coverings immediately and expose timbers to facilitate drying. Assess condition of timbers and replace where required. Where appropriate, treat the timbers. Ensure the timber is thoroughly dried before replacing finishes.
Solid or suspended concrete floors	Damage to sand/cement screeds insulation	Replace/repair with denser proprietary screed. Remove screeds and replace damaged insulation with closed cell type, repair with dense screed, sealed to wall.

considered are door boards, air vent covers, external skirting or inflatable membranes. In addition, having protection that is remote from the building, such as a flood gate, could also be included. Other external protection methods are available and many products are on the market. Specifiers need to be discerning about the choice or products and be sure that its limitations of use are known. Other issues to be considered are the need to quickly put into place the measures, but then be able to remove them immediately after a flood.

Table 9.2 illustrates the particular weaknesses associated with common forms of construction, and identifies possible repair options that are flood resilient. By adopting the preferred solutions for these forms of construction, subsequent flooding should be easier and less costly to repair, thus improving the whole life costs.

REFERENCES

Association of British Insurers (ABI), *Assessment of the Cost and Effect on Future Claims of Installing Flood Damage Resistant Measures*, May 2003.

Building Research Establishment (BRE), BRE Digest 410, *Cementitious renders for external walls*, CRC Ltd, Watford, 1995.

Building Research Establishment (BRE), Good Building Guide 18, *Choosing External rendering*, CRC Ltd, Watford, 1994.

Building Research Establishment (BRE), BRE Good Repair Guide 11, *Repairing Flood Damage* Part 1–4, CRC Ltd, Watford, 1997.

Building Research Establishment (BRE), BRE Report BR332, *Floors and flooring*, CRC Ltd, Watford, 1997.

Building Research Establishment (BRE), BRE Report BR352, *Walls, windows and doors*, CRC Ltd, Watford, 1998.

Building Research Establishment (BRE), BRE Report BR440, *Foundations, basements and external works*, CRC Ltd, Watford, 2000.

Building Research Establishment (BRE), BRE Report BR466, *Understanding dampness*, CRC Ltd, Watford, 2004.

British Standards Institution, BS5262, *Code of practice for external renderings*, BSI, London, 1991.

British Standards Institution, BS5492, *Code of practice for internal plastering*, BSI, London, 1990.

British Standards Institution, BS5628-3, *Code of practice for structural use of masonry*, BSI, London, 2001.

CIRIA, Flood products: *using flood protection products* – a guide for homeowners, CIRIA, London, May 2003.

Environment Agency, *Lessons learned from the autumn 2000 floods*, Environment Agency, London, 2001.

Foundation for the Built Environment (FBE), Report No. 2, *Climate Change and buildings*, FBE, Watford, UK, (editors – Phillipson, M. and Graves, H.), 2001.

Office of the Deputy Prime Minister (ODPM), *Preparing for Floods*, ODPM, London, 2002.

10

California Climate Change, Hydrologic Response, and Flood Forecasting

Norman Miller
Earth Sciences Division, Berkeley National Laboratory, Berkeley, California, USA

ABSTRACT: California has ecological, cultural, and geographical diversity with 10 natural bioregions (Northern Coastal, Central Coastal, Southern Coastal, Great Central Valley, Cascade Mountains, Sierra Nevada Mountains, East of Sierra Nevada, Mojave Desert, and Sonoran Desert), and several urban centers. Its population exceeds 33 million, with the largest concentrations in the Los Angeles Basin, San Francisco Bay Area, and within the Central Valley. It has an area of approximately 420,000 km^2, with over 2000 km of coastline, and 3000 km^2 of lakes, bays, and deltas. Areas within the San Francisco Bay Delta are below sea level and have levees, retaining walls, and drainage ponds, which along with the California coastal areas, are vulnerable to storm-generated tidal surges, flooding, erosion, and loss of property.

10.1 INTRODUCTION

California has experienced long dry and unusually wet periods. Oxygen-18 isotope analyses (Ingram et al. 1996) of San Francisco Bay sediments dating back to about 1200 A.D have indicated changes in salinity, which are indicators of fresh water inflow to the San Francisco Bay. Periods of very high fresh water inflow to the San Francisco Bay (i.e. increased precipitation and low salinity) alternate with periods of low fresh water inflow (i.e. decreased precipitation and high salinity). During the 1500s, there appears to have been a drought that lasted more than 50 years (Figure 10.1). Since the beginning of the 1900s, California has been wetter than average, and large engineering projects for water storage and conveyance of Sierra Nevada snowmelt runoff have contributed to the economic growth of California. The sustainability of this region is closely aligned with the availability of fresh water, the storage of this water during dry periods, and the reduction or adaptation to major floods that can potentially damage vital infrastructure.

California has a Mediterranean climate and receives precipitation from November to April, with May through October being dry month. Precipitation is from the Pacific Ocean and involves large-scale atmospheric circulations that are

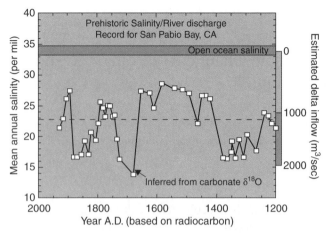

Figure 10.1. Oxygen-18 isotope dating of sediments indicates pre-historic periods of high
salinity and low inflow (low precipitation), and periods of low salinity and
high inflow (high precipitation). The red side-bar represents the estimated fresh
water inflow corresponding to the salinity concentration (Ingram et al., 1996).

influenced by Pacific Ocean sea surface temperatures (SSTs). The SST patterns
affect low frequency climate oscillations that impact precipitation rates in regions
around the world through natural teleconnected climate variations. For California,
the two most significant are the El Nino Southern Oscillation (ENSO) and the
Pacific multi-Decadal Oscillation (PDO). ENSO has a 3–5 year cycle, responds to
changes in the western equatorial Pacific Ocean SSTs, and is termed El Nino for
its positive phase, and La Nina for its negative phase. The PDO responds to SST
patterns in the northern, equatorial, and southern Pacific Ocean, has an approximately 30 year cycle, and is weaker than ENSO. Major El Nino phase events are
shown in Figure 10.2 and the global precipitation impacts due to the very large
1982–1983 El Nino are shown Figure 10.3.

During the last 50 years, there has been an observed global warming of the
lower atmosphere at rates greater than direct observations over the previous 150
years and greater than proxy studies that date back 1000 years (Figure 10.4). As much
as 30 percent of this recent warming can be attributed to natural climate variability, including ENSO, PDO, and other phenomena. However, studies have provided
strong evidence of climate change due to the emission of carbon dioxide (CO_2)
and other anthropogenic green house gas (GHG) emissions. Since the late 1950s,
observations have indicated that the global temperature of the lower atmosphere
(troposphere) has increased on average 0.1°C per decade, with the 1990s the
warmest decade of the century. Global input of CO_2 to the atmosphere from fossil
fuel combustion has exponentially increased since the beginning of the industrial
revolution (i.e. 1860) at about 4% per year. Atmospheric CO_2 measurements
(Keeling et al. 1989) from Mauna Loa, Hawaii beginning in the late 1950s show

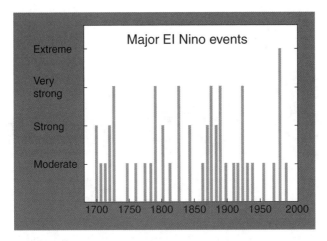

Figure 10.2. Major El Nino events since 1700. The 1982–1983 event was the largest on record.

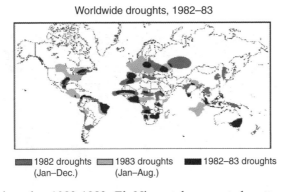

Figure 10.3. During the 1982-1983 El Nino, teleconnected patterns caused global impacts, most importantly droughts and floods.

increasing atmospheric CO_2 concentrations from near 280 ppm during the late 1950s to concentrations greater than 360 ppm in the 1990s (Figure 10.5). Ice core measurements from Vostok, Antarctica indicate that the atmospheric CO_2 concentration was less than 280 ppm for the last 100,000 years (Barnola et al. 1987). It is well understood that CO_2 and other anthropogenic emissions (e.g. methane, nitrous oxide, halocarbons) act as heat trapping GHGs, that radiatively force the global climate and increase tropospheric temperature. This increase in atmospheric CO_2 concentration and temperature appears to be causing a range of impacts, some of which may have adverse consequences.

In response to this warming and the potential impacts, the Intergovernmental Panel on Climate Change (IPCC) was established in 1988 and the First Assessment Report released in 1990 provided a discussion of these concerns and provided global and regional climate model projections, framed with uncertainties (IPCC

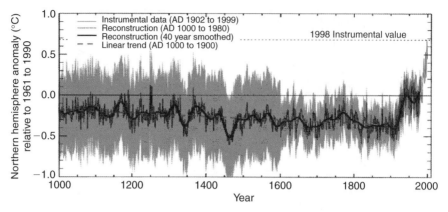

Figure 10.4. The observed temperature record from 1860 to present is shown in red, the reconstructed paleo-temperature record for the period 1000 to 1980 AD is shown in blue, with a 40 year smoothed version of the paleo-record in black. The linear trend (red dashed) decreases until about 1860, when historically high rates of temperature increase are observed.

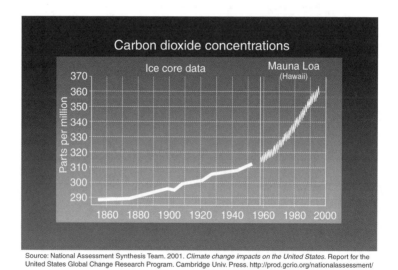

Source: National Assessment Synthesis Team. 2001. *Climate change impacts on the United States*. Report for the United States Global Change Research Program. Cambridge Univ. Press. http://prod.gcrio.org/nationalassessment/

Figure 10.5. The observed CO_2 concentration at Mauna Loa for 1959 to 1995 correlates well with the observed temperature increase shown in Fig. 4.

1990). In 1995, the IPCC Second Assessment Report (IPCC 1995) stated that the balance of evidence suggested a discernable human influence on global climate. The Third Assessment Report (IPCC 2001) has stated that there is new and stronger evidence that most of the warming observed over the last 50 years is attributable to human activities, and warming will continue throughout the 21st century.

The IPCC (2001) ranked the confidence limits of major impacts to water resources due to observed and projected climate change as very high (0.95–1.00), high (0.67–0.95), medium (0.33–0.67), low (0.05–0.33), and very low (0.00–0.05). There is high confidence that the timing and amount of runoff is changing, and very high confidence that watersheds with substantial snowpack will experience major changes as temperature continues to rise. The impacts of this trend are a decrease in available water resources in California, primarily during the summer months, and a potential increase in wintertime floods. There is high confidence that California's Sierra Nevada will experience a continued trend of decreased snow accumulation and earlier snowmelt (e.g. Lettenmaier and Gan 1990; Jeton et al. 1996; Miller et al. 1999; Wilby and Dettinger 2000; Knowles and Cayan 2002; Miller et al. 2003).

Precipitation change is very location-dependent and at present there is low confidence on changes in frequency, intensity, and persistence. There is medium confidence that there will be increased flooding in regions that experience large increases in precipitation. The IPCC (2001) suggests that flooding may become a major problem, raising concerns about property damage and levee failure. There is high confidence that increased temperature will cause sea level rise through thermal expansion and increased freshwater input (snow melt runoff, glacier calving). There is high confidence that this will impact groundwater aquifers, surface freshwater systems, and urban areas in The Netherlands, the San Francisco Bay-Delta, and other low lying regions. Sea level rise will cause an increase in erosion and storm surges increasing the extent of damage to coastal areas, bays, levees, and water conveyance structures.

10.2   MODELING CLIMATE CHANGE AND HYDROLOGY

Atmospheric-Ocean General Circulation Models (AOGCMs) represent a coarse spatial approximation of the global climate system and its response to changes in atmospheric composition. Climate projections are initialised by output from carbon cycle model projections indicating future CO_2 concentrations of 250 to 970 ppm representing GHG concentrations of 90 to 250% of the pre-industrial 280 ppm. This range brackets projected *Business as Usual* scenarios and significant GHG reduction scenarios. Aerosols such as sulphates and black carbon reflect solar radiation in the troposphere causing a cooling effect. Naturally occurring aerosols such as volcanic emissions also play into this uncertainty. Since the 1990s, these aerosols have begun to be added as part of emission scenarios models along with CO_2 concentrations.

The IPCC AOGCM projections of 21st century climate, with a transient increase in greenhouse gas emissions, suggest that the global mean near-surface air temperature will increase by 1.4 to 5.8°C, with a 95% probability interval of 1.7 to 4.9°C by 2100 (Wigley and Raper 2001). The potential for impacts on water resources

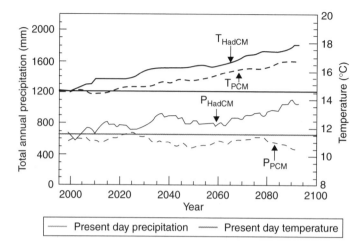

Figure 10.6. The California area-averaged temperature for HadCM2 and PCM show an increase from present day (red) of approximately 4.5°C and 2.1°C, respectively, by 2100. While the HadCM2 shows a precipitation increase of approximately 1.9% and the PCM decreases by 0.85% from the present day precipitation (green).

due to global warming requires a downscaled analysis of local watershed hydrology. To quantify the uncertainty in streamflow, upper and lower temperature and precipitation projections and specified incremental temperature and precipitation changes are used as input forcing to hydrologic models.

Miller et al. (2003) performed an analysis of California hydrologic response to 21st century future climate temperature and precipitation based on two AOGCM projections representing the spread of possible outcomes. The models used are the Hadley Centre HadCM2, a warm and wet projection, and the National Center for Atmospheric Research PCM, a cool and dry projection, relative to the mean of the IPCC AOGCM projections (Figure 10.6). Climate change *perturbations* of the projected watershed mean-area temperature and precipitation were derived from the temperature difference and precipitation ratio between the projected climatology and the simulated present day baseline climate (1961 to 1990).

10.3   HYDROLOGIC RESPONSE TO CLIMATE CHANGE

The hydrologic response was investigated using the U.S. National Weather Service Sacramento Soil Moisture Accounting Model and Anderson Snow Model (Burnash et al. 1973) for a set of representative study watersheds (Smith River, Sacramento River, Feather River, American River, Merced River, Kings River) forced by precipitation and temperature perturbations. These watersheds (Figure 10.7) were selected to evaluate the shifts in timing and magnitude for runoff in the Cascade

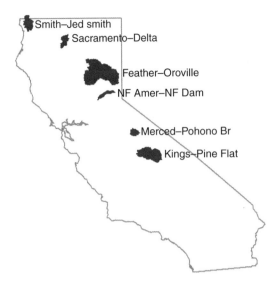

Figure 10.7. California study watersheds are the Smith River, Sacramento River, Feather River, North Fork American River, Merced River, and the Kings River. These were chosen to examine the runoff response for major California fresh water sources via similarity. The discussion focuses primarily on the Sacramento, American, and Merced Rivers.

Mountains, Sierra Nevada Mountains, and Northern Coastal mountains, as 80% of California's fresh water comes from these regions. Comparisons were made between present-day (1961–1990) hydrology and the hydrologic response to the relatively warm, wet projection and the cool, dry projection. Changes in the rain-to-snow ratio, streamflow, and the range in daily high streamflow return periods (i.e. change in probability of floods) were determined. The following discussion is focused on a subset of these watersheds; the Sacramento, the American, and the Merced.

The resulting HadCM2-forced peak flow increases for the Sacramento, American and Merced during 2010 to 2039. During 2050 to 2079 and the 2080 to 2099, the peak flow magnitude continues to increase, with the greatest increase at the American River (Figure 10.8, left column).

The peak flow timing during the 2080 to 2099 period for the American is a month earlier, occurring in February, while the Sacramento (rainfall-dominant watershed) timing remains relatively unchanged. The higher elevation Merced peak flow occurs one month later than the historical, and it also shows a secondary peak flow during December. This secondary high flow is due to increased early season snowmelt and higher snowline due to the increased temperature. The relatively cool, dry PCM-forced streamflow (right column figures) slightly decreases in total volume and significantly decreases during the March to July melt season. Peak flow remains close to the historical for the Sacramento and American for all projected periods, but the Merced shows an increase during the 2010 to 2039 period, and decreases during 2050

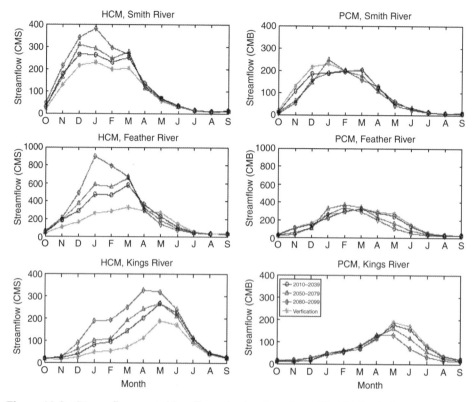

Figure 10.8. Streamflow monthly climatologies based on HadCM2 (left column) and PCM (right column), for the Sacramento River, American River, and Merced River, for present day (light blue, validation), 2010–2039 (dark blue, circle), 2050–2079 (green, triangle), and 2079–2099 (red, diamond).

to 2079 and 2080 to 2099. For these projections, the American shows an earlier peak flow of one month, while the peak flow for the other two watersheds shown here remains consistent with the historical peak flow timing. Under the scenarios studied, peak streamflow in the snowmelt runoff watersheds occurs earlier and with increased magnitude and decreased summer season flow.

From a water resources perspective, the most significant finding common to both the HadCM2 and the PCM is a 50 percent decrease in total snow for California by 2100 (Figure 10.9). The one exception is the very high elevation Merced, which is sufficiently cold that it remains below freezing during most of the snow accumulation season. Such reductions in available fresh water will impact agricultural and urban water use. April 1st is the time when the California Department of Water Resources determines the amount of water available for the dry season.

Changes in the snowmelt timing coupled with increased winter-time warm precipitation (rain) suggest an increased likelihood of high streamflow days that may result in floods. To evaluate likelihoods of future flooding, 30 year high flow days

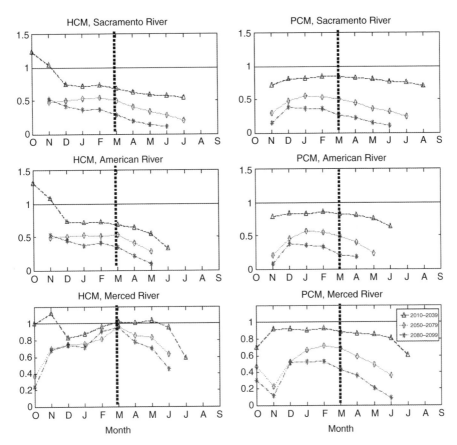

Figure 10.9. The ratio of projected to present day total snow for each month based on the HadCM2 (left column) and the PCM (right column) for the Sacramento River, American River, and Merced River, for present day (light blue, validation), 2010–2039 (dark blue, circle), 2050–2079 (green, triangle), and 2080–2099 (red, diamond). The vertical dashed line represents the end of season snow amount.

were calculated and ranked for both the HadCM2 and PCM, as well as for the present day period. Figure 10.10 indicates that for both the relatively warm, wet HadCM2 and cool, dry PCM, there is a significant increase in the likelihood of high flow days. For each curve shown in Figure 10.10, the median of the annual maximum daily flow (50%) increases with increasing temperature. The 5% exceedance high flow (the rightmost symbols in each plot) for the projected climates exceeds current conditions, implying an increased likelihood of high flow days (i.e. floods).

10.4   FLOOD MODELING

The ability to forecast high flows sufficiently far in advance under present and future climate conditions can reduce the loss of property and life. One of the first

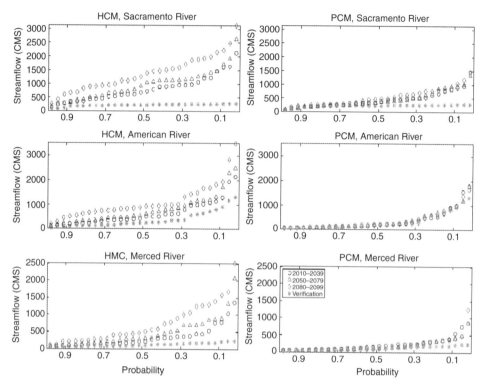

Figure 10.10. Daily streamflow excedance probability based on the HadCM2 (left column) and the PCM (right column) for the Sacramento River, American River, and Merced River, for present day (light blue, validation), 2010–2039 (dark blue, circle), 2050–2079 (green, triangle), and 2080–2099 (red, diamond).

successful 48-hour advance numerical weather and streamflow forecasts was performed in January 1995 (Miller and Kim 1996a) along the Russian River in Northern Coastal California (Figure 10.11). During the first week of January 1995, there were three storm systems that made landfall in the Northern Coastal region closely spaced in time. After the second storm, the ground became saturated and forecasts indicated that flooding (500 CMS) would occur 42 hours after the forecast was made. This simulation was based on a three-step procedure. Our modelling system (Miller and Kim, 1997) automatically obtained, via file transfer protocol (FTP), a global forecast from the U.S. National Weather Service. The global forecast provided our regional scale numerical weather prediction model with the initial and lateral boundary conditions needed to generate a fine-scale forecast, which in turn provided input to our coupled hydrologic model.

This initial work has since advanced with ensemble streamflow simulations and data assimilation. The U.S. National Weather Service-River Forecast Center has developed flash flood modelling techniques with increased forecast skill. There have been successful 72-hour streamflow forecasts when the weather system has

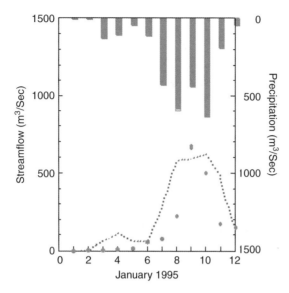

Figure 10.11. One of the first 48-hour numerical weather-streamflow predictions was made for the first 12 days of January 1995. The green dashed line is the streamflow forecast, the red dots are stream gauge observations, and the blue bars are the precipitation forecast (Miller and Kim 1996).

good predictability. Now-casting has also emerged as an accurate approach for evaluating weather systems at very short time intervals based on weather radar and remote sensed data.

During the last five years, the U.S. National Weather Service has been implementing the Advanced Hydrologic Prediction Services (AHPS). AHPS is a federally funded modernisation initiative that allows for an incremental introduction of new products and services across the U.S.. The AHPS provide new forecast products depicting the magnitude and uncertainty of occurrence for hydrologic events from hours to days to weeks. The AHPS leverages data and systems from the U.S. National Weather Service and collaborators, including Lawrence Berkeley National Laboratory. It consists of radar data, river gage data, weather observations, snow cover/melt data, precipitation forecasts, climate predictions, reservoir releases, and satellite data.

10.5   SUMMARY AND CONCLUSIONS

There is strong evidence that the lower atmosphere has been warming at an unprecedented rate during the last 50 years, and it is expected to further increase at least for the next 100 years. Warmer air mass implies a higher capacity to hold water vapor and an increased likelihood of an acceleration of the global water

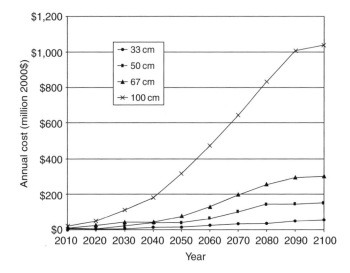

Figure 10.12. California costs for coastal protection for sea level rise of 33 cm, 50 cm, 67 cm, and 100 cm.

cycle. This acceleration is not validated and considerable new research has gone into understanding aspects of the water cycle (e.g. Miller et al. 2003). Several significant findings on the hydrologic response to climate change can be reported. It is well understood that the observed and expected warming is related to sea level rise. In a recent seminar at Lawrence Berkeley National Laboratory, James Hansen (Director of the Institute for Space Studies, National Aeronautics and Space Administration) stressed that a 1.25 Wm^{-2} increase in radiative forcing will lead to an increase in the near surface air temperature by 1°C. This small increase in temperature from 2000 levels is enough to cause very significant impacts to coasts. Maury Roos (Chief Hydrologist, California Department of Water Resources) has shown that a 0.3 m rise in sea level shifts the San Francisco Bay 100-year storm surge flood event to a 10-year event. Related coastal protection costs for California based on sea level rise are shown in Figure 10.12 (Wilkinson et al. 2002).

In addition to rising sea level, snowmelt-related streamflow represents a particular problem in California. Model studies have indicated that there will be approximately a 50% decrease in snow pack by 2100. This potential deficit must be fully recognised and plans need to be put in place well in advance. In addition, the warmer atmosphere can hold more water vapor and result in more intense warm winter-time precipitation events that result in flooding. During anticipated high flow, reservoirs need to release water to maintain their structural integrity. California is at risk of water shortages, floods, and related ecosystem stresses.

More research needs to be done to further improve our ability to forecast weather events at longer time scales. Seasonal predictions have been statistical and only recently have studies begun to use ensemble simulations and historical observations to constrain such predictions. Understanding the mechanisms of large-scale atmospheric dynamics and its local impacts remain topics of intensive research.

The ability to predict extreme events and provide policy makers with this information, along with climate change and hydrologic response information, will help to guide planning to form a more resilient infrastructure in the future.

ACKNOWLEDGEMENTS

Support for this manuscript provided by the US Department of Energy Office of Science, Office of Biological and Environmental Research under contract DE-AC03-76F00098. Climate change and hydrologic response research was supported by NASA Grant NS-2791 and California Energy Commission. The early flood forecasting model development and simulations were supported through a US DOE Laboratory Directed Research and Development Grant. This manuscript is Lawrence Berkeley National Laboratory Report LBNL-54041.

REFERENCES

J.M. Barnola, D. Raymond, Y.S. Korotkevitch, and C. Lorius, 1987: *Vostok ice core: A 160,000 year record of atmospheric CO₂*, Nature, 329, 408–414.

R.J. Burnash, R.L. Feral, and R.A. McQuire, 1973: *A generalized streamflow simulation system. In: Conceptualization modeling for digital computers.* U.S. National Weather Service.

B.L. Ingram, J.C. Ingle, and M.E. Conrad, 1996: *Isotopic records of Pre-historic salinity and river inflow in San Francisco Bay Estuary, in San Francisco Bay: The Ecosystem* (J.T. Hollibaugh, Ed.), Amer. Assoc. for the Advancement of Science, Pacific Division, San Francisco, CA, 35–61.

Intergovernmental Panel on Climate Change, 2001: *Climate Change 2001: The Scientific Basis.* Cambridge Univ. Press. 881pp.

Intergovernmental Panel on Climate Change, 1995: *Climate Change 1995: The Science of Climate Change.* Cambridge Univ. Press. 531pp.

Intergovernmental Panel on Climate Change, 1990: *Climate Change: The IPCC Scientific Assessment.* Cambridge Univ. Press. 362pp.

A.E. Jeton, M.D. Dettinger, and J.L. Smith, 1996: *Potential effects of climate change on streamflow: Eastern and western slopes of the Sierra Nevada, California and Nevada.* U.S. Geological Survey, Water Res. Investigations Rep. 95–4260, 44pp.

C.D. Keeling, R.B. Bacastow, A.F. Carter, S.C. Piper, T.P. Whorf, M. Heimann, W.G. Mook, and H. Roeloffzen, 1989: *A three dimensional model of atmospheric CO₂ transport based on observed winds: 1. Analysis of observational data.* In: Aspects of climate variability in the Pacific and Western Americas. Peterson (ed.), Geophysical Monograph, 55, AGU, 165–236.

N. Knowles, and D.R. Cayan, 2002. *Potential effects of global warming on the Sacramento/San Joaquin watershed and the San Francisco estuary. Geophys.* Research Letters, 29, 1891.

D.P. Lettenmaier, and T.Y. Gan, 1990: *Hydrologic sensitivities of the Sacramento-San Joaquin River Basin, California, to global warming.* Water Resources Res, 26, 69–86.

N.L. Miller, and J. Kim, 1996: *Numerical prediction of precipitation and river flow over the Russian River watershed during the January 1995 California storms.* Bulletin Amer. Meteorological Soc. 77, 101–105.

N.L. Miller, and J. Kim 1997: The Regional Climate System Model. *In Mission Earth: Modeling and Simulation for a Sustainable Global System.* Ed. M. Clymer and C. Mechoso, Soc. Comp. Sim. Inter., 55–60.

N.L. Miller, J. Kim, R.K. Hartman, and J. Farrara, 1999: *Downscaled climate and stream-flow study of the Southwestern United States.* J. Amer. Water Resources Assoc, 35, 1525–1537.

N.L. Miller, K.E. Bashford, E. Strem, 2003: *Potential impacts of climate change on California hydrology.* J. American Water Resources Association, 39, 771–784.

N.L. Miller, A.A.W. King, M.L. Wesely and others, 2003: The U.S. Department of Energy Water Cycle Pilot Study. Bulletin of the American Meteorological Society, In Review.

T.M.L. Wigley, and S.B.C. and Raper, 2001: *interpretation of high projections of global-mean warming.* Science, 293, 451–454.

R.L. Wilby, and M.D. Dettinger, 2000: *Streamflow changes in the Sierra Nevada, California, simulated using statistically downscaled general circulation model output.* In: Linking Climate Change to Land Surface Change, Ed. McLaren and Kniven, Kluwer Academic Pub., 99–121.

R. Wilkinson, and others, 2002: *Preparing for a climate change: California.* Report of the California Regional Assessment Group, U.S. Global Change Research Program. 850pp.

Conclusions

Andras Szöllösi-Nagy
UNESCO Natural Sciences Sector, Paris, France

Chris Zevenbergen
Dura Vermeer Business Development BV, Hoofddorp, The Netherlands

Over the last decades the world has witnessed a growing number of floods in urban areas. Climate change and rapid urbanisation will exacerbate this trend. Flooding incidents in urbanised catchments can lead to much public concern and anxiety; the economical impact is severe. Non-structural approaches for urban flood mitigation have been adopted as structural approaches have proven to offer only partial solutions. Urban planning guidelines and flood management strategies and tools should, therefore, be part of an integral approach to the problem.

New approaches to the accommodation of floods are needed in order to create robust and sustainable solutions that can cope with the ever-increasing urban pressure on flood-prone areas and the uncertainties created by climate change. Keeping water out of urban areas is often not the perfect solution; accepting and preparing for some degree of flooding will in many cases be a more sensible solution, from technical and financial perspectives, but also from social and environmental perspectives. External factors can also influence the choice between total avoidance of flood or a design for controlled flooding. Measures to reduce flooding may ultimately lead to increased flood risks elsewhere. Decision makers need to assess the risk as well as the cost and benefit of alternative strategies addressing different spatial scales such as zoning controls, regulation of construction on flood plains, flood proofing, flood insurance, embankments, flood diversion channels and real-time monitoring. At present, there are no appropriate tools available for an integral assessment of these risks.

The challenges created by the recent surge in floods and flood-related damage are of such magnitude that individual countries do not always have the capacity and financial resources to develop new tools, guidelines, techniques, instruments, materials, products and services on their own. This certainly holds true for developing countries, where the urban poor are often forced to settle in flood-prone areas and lack the adaptive capacity to cope with flood events. Urban floods cannot be viewed, therefore, as a problem for individual countries. At present some countries are more proactive in managing urban floods and the opportunity exists for information exchange alongside the need to develop assessment tools to implement, facilitate and enhance urban flood adaptation strategies. Each country and even each city has a set of particular conditions and problems which precludes the automatic application of imported solutions. Nevertheless, as shown in the preceding chapters, a wealth of experience and information on, for instance, the incorporation of risk-reduction measures in urban planning, training and educational programs for decision makers and community leaders, empowerment and participation of civil society and communities programs and enforcement of construction standards have been accumulated by countries facing urban floods. This wide-ranging body of knowledge needs to be shared in order to learn how it may benefit others. Existing international networks, such as The International Network of Cities (UNESCO/IHP and the Academie de l'Eau), the European network Spid'O on Spatial Development and Water Management (Netherlands Water Partnership) and IFNet (International Network on Flood Management, which was launched at WWF3), should be exploited to promote and facilitate this exchange.

On the basis of the recommendations of the preceding chapters on future needs, at least three interrelated strategic areas can be identified which require more research:

− Urban planning: in general flood mitigation receives little attention in the spatial and development planning process. The information available on best practices is limited or even absent to practitioners, and politicians are often reluctant to take measures necessary to mitigate the impacts of urban floods. Mechanisms for increasing the capacity of local and state authorities and non-governmental organisations to analyse urban flood impacts and formulate strategies should be developed and implemented.

− Vulnerability assessment: the establishment of efficient flood management strategies includes comparative assessments of the potential financial benefit of flood management scenarios in terms of avoidance of damage. Damage assessments can be made using tools that relate to the modelled flood levels and extend to the property affected by the flood. Due to a lack of data and analysis of the technical performance and economics of implementation and maintenance the formalisation and encouragement of flood-proofing techniques and sustainable urban drainage systems are curtailed. Research is also needed to document the outcome of flood management plans, gains in reduced vulnerability and the effectiveness of the planning process itself.

− Perception of risk: the relationship between the perception and handling of risks and the consequences of urban floods must be better understood, because the response to risk also determines the potential for damage reduction.

Author Index